The Management of Water Quality and Irrigation Technologies

Edited by Jose Albiac and Ariel Dinar

from Routledge

First published by Earthscan in the UK and USA in 2009

Copyright © Jose Albiac and Ariel Dinar, 2009

This book is the outcome of the International Expo 'Water and Sustainable Development' held in Zaragoza, Spain, in 2008. Support from the Spanish Ministry of Environment, Caja Rioja, Government of Aragon and the World Bank is acknowledged.

ISBN: 978-1-84407-670-3 (hbk)
ISBN: 978-0-41584-939-5 (pbk)

Typeset by MapSet Ltd, Gateshead, UK

Cover design by Susanne Harris

For a full list of publications please contact:

Earthscan

2 Park Square, Milton Park, Abingdon, Oxon OX14 4RN

711 Third Avenue, New York, NY 10017

Earthscan publishes in association with the International Institute for Environment and Development

A catalogue record for this book is available from the British Library

Library of Congress Cataloging-in-Publication Data

Albiac, José.
    The management of water quality and irrigation technologies / José Albiac and Ariel Dinar.
        p. cm.
    Includes bibliographical references.
    ISBN 978-1-84407-670-3 (hardback)
    1. Water quality management. 2. Irrigation engineering. I. Dinar, Ariel, 1947– II. Title.
    TD365.A75 2008
    363.739'46–dc22

                                                                        2008034728

Contents

Part I Non-point Source Pollution Regulation Approaches

Part II Irrigation Technology to Achieve Water Conservation

List of Figures, Tables and Boxes

Figures

Tables

Boxes

List of Contributors

Jose Albiac is research fellow at the Agrifood Research and Technology Centre (CITA-DGA, Spain). He is an economist working on water resources management and forest economics.

Llorenç Avellà is professor of agricultural and environmental economics at the Universidad Politécnica de Valencia and a member of the Centro Valenciano de Estudios del Riego (UPV). His work focuses on land use change and water economics and policy. He has participated in 49 research projects and has been a consultant to several governments. He is the author of about 40 articles and 15 books.

Richard Batiuk is the associate director of science at EPA Chesapeake Bay Program Office (CBPO).

Elena Calvo is associate professor at the Department of Economic Analysis, University of Zaragoza (Spain). She is a mathematician working on mathematical modelling and simulation of environmental systems.

Baryohay Davidoff is chief of the Agricultural Water Management & Financial Assistance Unit, California Department of Water Resources, in Sacramento. His work includes a wide range of water use efficiency and water resource management issues, such as developing and implementing programmatic, technical and financial assistance, and loans and grant programmes. For the past 20 years, Dr Davidoff has worked on water resources management and efficiency processes, including water management planning and implementation, agricultural drainage and the California Water Plan Updates. These efforts include on-farm, local, regional and state-wide level processes. Dr Davidoff is a board member and served as president of the California Irrigation Institute.

Ariel Dinar is a professor of environmental economics and policy, and director of the Water Science and Policy Center, University of California, Riverside. This book was prepared while he was lead economist of the Development Research Group at the World Bank. His research focuses on international water and cooperation, approaches to stable water allocation agreements, water and climate change, economics of water quantity/quality, and economic aspects of policy interventions and institutional reforms. His most recent undertaking is the Resources for the Future (RFF) book series *Issues in Water Resource Policy* that aims to produce publications on contemporary water policy issues in various countries and states.

Marta García-Mollá is an associate professor at the Universidad Politécnica de Valencia and a member of the UPV. She has worked on research projects in the area of water policy and planning and in research projects related to the economics of water in agriculture. Her PhD is on the effects of an increase in the cost of water consumption in Mediterranean agriculture.

Munther Haddadin served as CEO and chairman of the Jordan Valley Authority (1982–1987), as senior member of the Jordan delegations to the Middle East Peace Process (1991–1995), as Minister of Water and Irrigation of Jordan (1997–98) and is a courtesy professor at Oregon State University (2001–present), University of Oklahoma (2003–present) and at University of Central Florida (2006–present). He is the author and co-author of several books and articles.

Lewis Linker is the Chesapeake Bay Program modelling coordinator, and works with colleagues throughout the Chesapeake Bay Program to develop linked models of the airshed, watershed, estuary and living resources of the Chesapeake. He is an environmental scientist at the EPA Chesapeake Bay Program Office (CBPO).

Mithat Mema is vice chancellor of the Aleksander Moisiu University (Albania) and chairman of the Department of Economic Analysis. He is an economist working on environmental economics, with extensive experience in non-point source pollution.

Edwin Ongley was formerly a director in the National Water Research Institute of Environment Canada (retired) and former director of the United Nations GEMS/Water Programme. He has worked in China for 20 years on a variety of international water and environment management projects at national, provincial and local levels.

Chris Perry worked for the World Bank for more than 20 years, primarily on water-related projects in south Asia. He was seconded to the International Water Management Institute for five years, serving as deputy director general. He is now a freelance consultant with a particular interest in the role of remote sensing in improving water resources management.

Marc Ribaudo is currently an agricultural economist with the Economic Research Service (ERS). He has been working at ERS since 1983. He is currently in the Resource, Environmental and Science Policy Branch of the Resource and Rural Economics Division. He has worked primarily on water quality issues, including the water quality impacts of agricultural and conservation programmes, the impacts on agriculture of policies for protecting water quality, and the design of policies for reducing non-point source pollution. He is currently leader of a project examining the role private markets for environmental services may play in supporting conservation measures on farms.

Gary Shenk is an environmental scientist at the EPA CBPO.

Yu Tao worked for seven years in the Survey and Design Institute of Water Conservancy and Hydro-Electric Power in Xinjiang, China. He received his PhD from Peking University and is currently a research scientist in the Chinese Research Academy of Environmental Sciences, Beijing.

Ping Wang is a senior research scientist at the University of Maryland Center for Environmental Science.

Yi Wang is a professor and deputy director-general at the Institute of Policy and Management, Chinese Academy of Sciences. His research has been focused on strategic issues and public policy studies on sustainable development. He is a standing member of the Board of Directors of several academic societies. He has also been a senior adviser to several international organizations and foundations.

Michael Young holds a research chair in water economics and management at the University of Adelaide and is a fellow of the Academy of Social Sciences in Australia. Awarded the Land and Water Australia Eureka Award for Water Research in 2006, Mike spent 30 years with the Commonwealth Scientific and Industrial Research Organisation (CSIRO) Land and Water and established its Policy and Economic Research Unit. In 2005, he was awarded a Centenary Medal for his contribution to environmental economics.

List of Acronyms and Abbreviations

ACC	Agricultural Credit Corporation
ADB	Asian Development Bank
AF	acre-feet
AFO	animal feeding operation
ARS	Agricultural Research Service
ASAE	American Society of Agricultural Engineers
ASCE	American Society of Civil Engineers
BMP	best management practice
BOD	biochemical oxygen demand
CAFO	concentrated animal feeding operations
CAP	Common Agricultural Policy
CAT	climate assessment tool
CBMP	Chesapeake Bay Management Program
CBP	Chesapeake Bay Program
CIHEAM	Centre International des Hautes Etudes Agronomiques Méditerranéennes
CIMIS	California Irrigation Management Information System
CMAQ	community multiscale air quality
COAG	Council of Australian Government
COD	chemical oxygen demand
CSO	combined sewer overflow
CSP	Conservation Security Program
CVER	Centro Valenciano de Estudios de Riego
CWA	Clean Water Act
CWSRF	Clean Water State Revolving Fund
CZARA	Coastal Zone Management Act Reauthorization Amendments
DU	distribution uniformity
DWR	Department of Water Resources
EC	electrical conductivity
EGU	electric generating unit
EIA	Environmental Impact Assessment
EII	Environmental Improvement Initiative
EMC	event mean concentration

EPB	Environmental Protection Bureau
EQIP	Environmental Quality Incentive Program
ESYRCE	survey on crop acreage and yields
ET	evapotranspiration
ETo	reference evapotranspiration
EWMP	Efficient Water Management Practice
FIFRA	Federal Insecticide, Fungicide and Rodenticide Act
FY	fiscal year
GDP	gross domestic product
GIS	geographic information system
HLWPR	High Level Water Policy Reform
HRU	hydrological response unit
IAMZ	Instituto Agronómico Mediterráneo de Zaragoza
IBRD	International Bank for Reconstruction and Development
ICID	International Commission on Irrigation and Drainage
IE	irrigation efficiency
ILRI	International Institute for Land Reclamation and Improvement
INE	Instituto Nacional de Estadística
IRYDA	National Institute for Agricultural Reform and Development (Spain)
ITRC	Irrigation Training and Research Center
IWMI	International Water Management Institute
LR	leaching requirement
LULUF	land use, land use change and forestry
mcm	million cubic metres
MEP	Ministry of Environmental Protection
MS4s	municipal separate stormwater sewer systems
MWR	Ministry of Water Resources
NADP	National Atmospheric Deposition Program
NAP	National Action Plan
NCC	National Competition Council
NCP	National Competition Policy
NHT	Natural Heritage Trust
NIP	National Irrigation Plan
$N\text{-}NO_3^-$	nitrogen in nitrate (1mg $N\text{-}NO_3^- = 4.43$mg NO_3^-)
NPDES	National Pollutant Discharge Elimination System
NPS	non-point source
NRD	Natural Resources District
O&M	operation and maintenance
OECD	Organisation for Economic Co-operation and Development
PAICAS	Integral Use Plan for the Segura Basin (Spain)
ppm	parts per million
RAP	Rapid Appraisal Process

SAV	submerged aquatic vegetation
SDWA	Safe Drinking Water Act
SEPA	State Environmental Protection Administration
TMDL	total maximum daily load
TN	total nitrogen
TP	total phosphorus
USAID	United States Agency for International Development
USDA	United States Department of Agriculture
USEPA	United States Environmental Protection Agency
USGS	United States Geological Survey
WATECO	Working Group on Water Economics of the Common Implementation Strategy
WFD	Water Framework Directive
WMP	Water Management Plan
WQA	Water Quality Act
WQSTM	Water Quality and Sediment Transport Model
WUA	water user association

1

Introduction

Jose Albiac and Ariel Dinar

This book is an outcome of the conference 'Water: Economics, Policy, Politics and Agricultural Celebration' held in Zaragoza (Spain) at the beginning of the International Expo ZH2O 'Water and Sustainable Development' in 2008.[1] Two important questions examined in the conference were non-point pollution and irrigation technologies for water conservation. The book describes experiences from several countries that highlight both achievements and failures in dealing with pollution and irrigation issues. These topics are presented here taking into account the institutional, policy and technological circumstances in different countries and regions.

All over the world, the pressure on water resources is mounting because of the ever-increasing growth in population and economic activities.[2] As a result, the environmental sustainability of aquatic ecosystems in most regions around the world seems to be more and more difficult to achieve. One optimistic approach to addressing the problem is to design public policies and market institutions that will spur new technologies able to augment income and wealth, and reduce human impacts on water resources. Another more sombre approach is to limit or reduce water extractions and pollution loads as part of adopting more austere lifestyles necessary for ecosystem conservation. The key issue is whether knowledge, policies and technological advances will be able to avoid ecological disaster, without the adoption of the Malthusian approach based on curbing income and population.

Water is mostly an impure public good or common pool resource, characterized by non-exclusion (no exclusive access) and rivalry in consumption (consumption reduces the available quantity); whereas a private good is characterized by exclusion and rivalry in consumption.[3] Typical examples of common pool resources are irrigation, fish stocks and forests.

The Dublin Declaration of 1992 indicates in its fourth point that water should be considered an economic good, in order not only to improve water use efficiency and equity, but also to attain conservation and protection. This is the approach taken by many countries, and in particular by the European Union in its Water Framework Directive, which promotes water pricing as the solution for water use efficiency, conservation and protection.

The problem with this approach is that the price mechanism can work only where water is a private good traded in markets. An additional problem is that markets face difficulties when dealing with environmental externalities, and cannot guarantee water resource conservation and protection. Urban and industrial uses have the characteristics of using water as a private good, but irrigation is different because it has the characteristics of using water as an impure public good and also has environmental externalities. An aquifer is a good example of a common pool resource, with rivalry in consumption but non-exclusion,[4] and environmental externalities from overdraft or degradation of its quality.

Water pricing could modify water consumption where markets exist, such as for connected urban and industrial uses, but not for agricultural or environmental uses. The cases of California and Australia (Chapter 6) seem to demonstrate that water markets are unable to internalize the environmental externalities.

The protection and conservation of water resources that are common pool resources requires cooperation by agents and collective action. To understand the problem of common pool resources and collective action, the example of pollution is a case in point. Pollution abatement can be undertaken by agents without cooperating with other agents, but this leads to an insufficient level of abatement. When there is cooperation among agents, abatement increases up to the level which is optimum for society satisfying the condition of efficient provision of public goods.

The sustainable management of water resources requires the availability of accurate information on the economic value of the services provided by the water resources and their associated ecosystems. The values of these environmental services are needed in order to find out the optimal level in the objectives of water policies, or the thresholds for cost-efficient measures when optimal levels are unknown and reasonable thresholds are sought. Information on economic valuation of environmental services is quite scarce in international literature, although the global value of these services seems quite high (Freeman, 1979, 1990; Loomis, 1997).

The purpose of this book is to show the water resource situation in some of the countries that have significant water scarcity and water quality degradation problems. The different policy measures undertaken by these countries are examined in the following chapters, taking into account the institutional settings and the requirements and availability of information and biophysical knowledge. Then the outcomes from pollution and irrigation policies are evaluated, and the general finding across countries is that both water quality and water scarcity policy goals are quite difficult to achieve with current policies. These difficulties seem to call for a revamp of the policy mix and the institutional setting in order to move towards a sustainable use of water resources.

The book is structured in two parts: Part I presents chapters on pollution and water quality in China, the US, the European Union, Spain and Australia, and Part II presents chapters on irrigation technologies in Jordan, California, Spain and Australia.

Part I Non-point source pollution regulation approaches

Good water quality is an essential condition for having living rivers with healthy aquatic ecosystems. At present, the pressure on water resources is growing rapidly both in terms of expanding water extractions and quality degradation from pollution loads. Water quality degradation is pervasive in most water-courses around the world, driven by the escalating pollution loads from anthropogenic point and non-point sources.

In high-income countries, there have been large investments in sewage networks and water treatment facilities during recent decades to control point pollution, which have stabilized or in some cases reduced the concentration of pollutants in rivers. Non-point pollution is much more difficult to tackle, because control measures are very difficult to design, implement and enforce. As a consequence of the abatement of point pollution, the relative importance of non-point pollution loads is increasing in high-income countries. In medium- and low-income countries, rivers and aquatic ecosystems are being degraded by the surge in point pollution loads from urban and industrial sources, and large tracts of watercourses have become unsuitable for many water uses.

Part I starts with the chapter by Ongley and Tao on the problems of assess-ing non-point pollution in China, and the influence of miscalculation on policies and regulations. Non-point pollution has become a critical policy issue in China in recent years, despite being a new field of study in China. The chapter starts by reviewing the five estimation methods of non-point pollution frequently used in China. One immediate conclusion is that there is a problem of overestimation of pollution loads when using these methods. This problem derives from the biophysical differences between China and the US, where these methods were first developed.

The key message from Ongley and Tao is that research results are influenc-ing pollution control policies to an improper level. Research on non-point pollution started in the 1980s in China, but there has never been a systematic appraisal of non-point pollution impacts. Models of non-point pollution in the US are empirically based on large amounts of real data, but the application of these techniques in China is hampered by the lack of the empirical databases necessary for calibration. In China, where point sources are not controlled, the range of non-point pollution estimates are too wide, especially for large basins. The question is important because non-point source estimates are starting to influence public policy, and the design of measures for pollution control. In some important wetlands and lakes, the control of eutrophication may require control of non-point sources. However, the best approach to pollution reduction should remain focused on point sources. In addition, the gap between policy and legislation, and enforcement has to be reduced, which is a common problem in most countries.

The chapter by Linker et al describes how the eutrophication problem in Chesapeake Bay has been addressed by a partnership of state and federal governments. The learning process of the Chesapeake Bay Program (CBP) has been supported by modelling, monitoring and research, which are the three key elements of the programme. The CBP modelling effort was a response to the questions put forward by decision makers, namely:

1   What are the levels of nutrients and sediments from the watershed and airshed needed to reach desired water quality standards?
2   What is the magnitude of point and non-point pollution loads?
3   What are the reductions of pollution to be made and their breakdown by media (water, air, land) and state, that are cost efficient and equitable?

Six states and the federal government agreed to pollution reductions with cap loads by tributary and jurisdiction. These cap loads drive the measures taken by each state to reduce pollution and meet the thresholds. Monitoring data underlie all integrated models, and model outcomes match the observed data. Monitoring programmes require support from modelling to explain observations. The CBP integrated models will be used to examine the measures needed to attain the water quality standards desired by 2020 and 2030, and these decisions will influence plans at federal, state and local levels in the coming years. The CBP voluntary approach is working since pollution loads have been reduced; however, a more regulatory approach could be necessary for further reductions.

In Chapter 4, Albiac et al present the water resources management situation in Spain and Europe, where water scarcity and quality degradation are important issues. Water quality degradation remains high in many river basins in both Spain and Europe, despite a comprehensive body of regulations and substantial investments in wastewater treatment plants. In Spain, some policies such as the National Irrigation Plan and the Water Quality Plan have the potential to address scarcity and quality degradation, but other water policies seem misguided.

Both the Spanish government and the European Commission advocate water pricing in irrigation, and the use of the Common Agricultural Policy (CAP) for cross-compliance, which seem to be flawed policy options.[5] The investments in advanced irrigation technologies undertaken by the Spanish government are much more promising than the European Water Framework Directive approach based on water pricing and pollution limits.

The regulation approaches to non-point source pollution in the US are described in Chapter 5 by Ribaudo. In the US, agriculture is the leading source of water quality impairment in rivers and lakes, and a major source in estuaries. Different incentives have been implemented at federal and state levels to reduce non-point pollution loads from farms, and the usual approach taken in the US is through voluntary mechanisms.

Some enforceable policy instruments are used for non-point pollution, such as cross-compliance, technology standards, performance standards and taxes, and emissions trading. However, the use of enforceable mechanisms for non-point pollution control has been the policy of last resort, and their use is more widespread at state than at federal level. The more restrictive programmes are directed at serious problems where voluntary approaches have failed. It seems that cost-effective non-point pollution control requires a policy framework that uses a combination of tools designed for specific watersheds.

In Chapter 6, Young analyses Australian experiences of controlling non-point pollution. Australia makes extensive use of market-based instruments to improve water use, and there are plans under way to apply these instruments to the control of non-point pollution.

There is a salinity trading arrangement between four states in the Murray–Darling basin, where states get salinity credits and debits, linked to investments to reduce salinity. Another example of market-based instruments is the dryland salinity control in the Bet Bet catchment in Victoria, where there is a cap and trade programme for salinity, including individual and collective payments if the desired outcome is reached. Victoria is also using a combination of water trading and charging to reduce salinity, because water trading has caused non-point pollution problems. The scheme is based on a zoning system with charges among zones and prohibition of trade to the zones with more severe problems of salinity. South Australia has a salinity offset trading programme, where new irrigation developments have to offset their salinity impacts from reductions elsewhere, for example, decommissioned irrigation areas. The key message is that market-based approaches to non-point pollution are more flexible and foster innovation, and therefore are more cost efficient.

Chapter 7 by Wang completes the presentations in Part I on water quality policies. China is facing acute problems of water scarcity, water pollution, degradation of aquatic ecosystems, and more frequent extreme events. The larger threat at present is water pollution, making changes in water management unavoidable. Water quality degradation is caused by the enormous increase in point pollution loads from industries and urban centres, and also from agricultural non-point pollution. These pollution loads make 30 per cent of river tracts checked unsuitable for industrial use or irrigation, and the effects are quite serious for the population because 300 million people do not have access to safe drinking water.

There is a water management failure in China, because the solutions have been based on technological and engineering measures, while the institutional and management arrangements have been neglected. The pollution problem is compounded by the low efficiency of water use in irrigation, industries and urban centres. Water governance should be improved by taking an integrated management approach based on stakeholders' involvement in management through cooperation, strong basin authorities, coordination of administrative bodies, laws and regulations that are truly enforced, and appropriate economic instruments and incentives. The importance of particular aquatic ecosystems should be evaluated to set priorities of protection for rivers and river sections.

Part II Irrigation technology to achieve water conservation

Water scarcity is becoming a serious problem in many arid and semi-arid regions of the world, where irrigation is the primary use of water. Worldwide extractions for the 280 million hectares of irrigated land amount to 2300km^3 per year. This amount represents 70 per cent of total water extractions.

Collective irrigation is based on dams and channel networks, and social control over extractions can be implemented quite easily, provided that the institutions are in place and the policy decision is taken and supported by stakeholders. However, the problem of overdraft is much more difficult to solve in the case of aquifers with individual pumping, because aquifers are common pool resources and their control entails cooperation from all the agents managing the wells, requiring a much more complex institutional setting and quite elaborate decision-making processes based on trust.

Irrigation efficiency varies between countries, influenced by both technology use and policies supporting the adoption of advanced irrigation technologies. A common policy response to water scarcity in regions with large irrigation acreage is to promote advanced irrigation technologies. This is the case in countries that lead in irrigation technology adoption and supporting policies, because high scarcity pushes them to find innovative solutions, both technical and policy related. The recent multibillion dollar investments in irrigation technologies in Spain, and those proposed in Australia, are examples of the importance that both countries place on irrigation technologies. The chapters on irrigation technology share the experiences in facing water scarcity and drawing up the necessary policy for technology development and adoption.

The concept of efficient use of water is quite popular among experts and decision makers as the best way to solve water scarcity problems. But efficiency gains do not translate necessarily into water savings or reduced extractions from watercourses. One example is the urban sector connected to sewage networks, where almost all water returns to treatment plants for reuse. Increasing the efficiency of household water devices and usages does not save much water. Another example is the efficiency gained by adopting advanced irrigation technologies, because these efficiency gains at plot level are usually coupled with more evapotranspiration at watershed level, driven by more demanding crops or by expanded irrigation acreage. Introduction of advanced irrigation technologies in areas with high water scarcity seem to lead to more evapotranspiration and less return flows, and the consequent decrease in the water flows running in rivers and watercourses in the basin, which are needed to support aquatic ecosystems.

In Chapter 8, Haddadin describes the process by which advanced irrigation technologies were introduced in Jordan, and the factors and policies encouraging adoption. Scarcity was the main driver, together with the search for gains in water efficiency and crop yields. Advanced technologies were adopted subsequently in the Jordan Plateau based on individual aquifer pumping, but this led to overdraft and efforts to curb excessive extractions. Jordanian farmers seem to have financial problems at present because of the elimination of subsidies and the rise in input costs, and they also have to face a dwindling supply of water and labour.

Davidoff presents in Chapter 9 the three essential elements to enhance irrigation efficiency and water conservation as they have been applied in California in response to rising water scarcity. The first element is availability of information on crop water use and consumption; the second is optimization of irrigation systems to attain water distribution uniformity in plots; and the third is that water should be available when needed. The chapter describes the experience in California of using public policy and technical and financial assistance to advance the three elements, through legislation requiring users to implement water management plans and efficient water management practices. The cost of implementing efficiency recommendations in California during the next 25 years could be up to US$4 billion, although in the past five years public funds to support agricultural efficiency investments have amounted to only US$80 million.

Perry reviews in Chapter 10 the debate on efficiency, pointing out that the traditional engineering concept of efficiency at field scale is not very useful at basin scale. The difficulties arise because some sectors consume the water, whereas others return almost all their water to the system. The traditional terminology of efficiency, saving, waste and loss is quite confusing when moving to basin scale. An example cited by Perry is household devices that may reduce water diversions, water treatment costs, water storage upstream, and dewatering of river tracts; but no water 'savings' can be made with these devices. A new terminology is needed to describe water use in terms of water balance in the basin. Withdrawals should be classified in consumed fraction (beneficial and non-beneficial) and non-consumed fraction (recoverable and non-recoverable). Then scarcity can be offset only by curbing non-beneficial consumption or non-recoverable flows.

Under this analytical framework, advanced irrigation technologies are likely to increase water consumption and reduce aquifer recharge and return flows, leaving less water for ecosystems. Perry concludes that economic instruments such as water pricing and water markets should be analysed under this new framework to avoid undesired outcomes from water pricing and water markets leading to more scarcity by higher extractions and consumption.

Avellà and García-Mollá analyse in Chapter 11 the institutional setting of irrigation in Spain, and the recent process of technology adoption. Lack of coordination among federal, state and local public administrations has triggered serious conflicts aggravated by water politics to gain votes. Another pervasive problem in water management in Spain is the lack of an effective control over concessions, and this failure has been driving an enormous expansion in water demand. The National Irrigation Plan was launched by the government to reduce water scarcity and pollution emissions. But preliminary results seem to indicate that water savings from the large investments have been used for acreage expansion or in more water-demanding crops.

The factors explaining modernization depend on farm characteristics. Large farms that rely on surface or subsurface water adopt technologies to achieve economies of scale, reduce costs and enhance product quality. Medium-sized and small farms based on subsurface water adopt technologies to expand acreage or reduce overdraft in order to lessen salinity problems. Medium-sized and small farms with flood irrigation only adopt technologies because of the subsidies incentive from the public administration. The final message from the chapter is the following: if water authorities want water savings in plots and network channels to translate into savings for whole watersheds and districts, they have to control changes in irrigation acreage and crop mix, or control the water balance of extractions and returns.

Chapter 12 by Young presents the case of Australia, where market-based instruments such as water trading have been introduced to improve water use. The question examined is whether the important process of irrigation technology adoption is being driven by the Australian water reform. Young shows that the introduction of water trading, water pricing and administrative separation seem to favour technological adoption, while markets move water to where new technologies are adopted. However, the process requires having in place a robust entitlement and allocation system.

Young indicates that there has been considerable progress in adopting efficient technologies, although less progress on over-allocation because markets facilitate individual changes but not collective changes. The key message from the chapter is that market-based instruments, combined with more conventional approaches, are able to achieve better control on water resources at less cost to the government and the community.

Lessons learned and policy implications

Improvement in the management of water resources in both developed and developing countries requires better information and knowledge about surface and subsurface resources, and their associated ecosystems. The examples of the

Chesapeake Bay in Chapter 3, non-point pollution in China in Chapter 2, and the European Water Framework Directive in Chapter 4, show that information and knowledge are essential ingredients for reasonable water policies.

Knowledge of the underlying biophysical processes is also a key aspect for water management, especially for aquifer management and non-point pollution control, where information is needed on aquifer characteristics and dynamics, and on pollution at local and watershed scales. Regarding pollution, the information should cover emission loads at the source, the pollutants' transport and fate processes, and ambient pollution in watercourses. Additionally, the absence of economic valuation of damage to aquatic ecosystems from aquifer overdraft and non-point pollution, precludes evaluating the benefits from policy measures. The lack of basic information and biophysical knowledge reinforces the strategic behaviour by stakeholders and states in federal countries, and makes the entire process of designing, implementing and enforcing reasonable and effective policy measures very difficult.

The European Water Framework Directive is a good example of strategic behaviour by countries and basins, as this strategic behaviour is already happening with the rest of environmental and non-environmental policies in the European Union.[6] European countries and basins have ample room to manoeuvre in each phase of the Water Framework Directive:

- the type of description of pressures, impacts and economic analysis;
- the kind of data taken in water measuring networks;
- the classification of water bodies;
- the water bodies that are considered as heavily modified (not requiring improvement); and
- the declaration of disproportionate costs (which let countries put aside measures). Countries have also ample slack in the design, implementation and enforcement of policy measures.

Besides information and knowledge, another essential feature for competent water management is the right institutional setting. One component of this institutional setting is to have strong basin authorities rather than the state or provincial authorities found in the US, Australia and China. The second equally important component is the involvement of stakeholders in basin authorities, so that all decisions in each watershed are made and respected by the stakeholders' representatives. The chapters of this book on water quality and irrigation technologies for water conservation demonstrate that successful water policies are the ones supported and carried out by stakeholders. In the absence of stakeholders taking the decisions, water policies are doomed to disaster.

A final issue worth mentioning is the revision of the popular concept of the efficient use of water among politicians and other decision makers. Both Perry in Chapter 10 and Avellà and García-Mollá in Chapter 11, analyse the validity of claims about water savings from efficiency gains in both the irrigation and urban sectors. The issue is important because Spain has just spent US$10 billion in irrigation technologies over the past few years, and Australia is preparing to spend US$7 billion in the coming years. Both are investing these large sums mainly to lessen water scarcity, but to discover whether they are correct in this undertaking will necessitate further enquiry.

Notes

1   The conference received support from the Spanish Ministry of Agriculture, Caja Rioja, Confederación Hidrográfica del Ebro, Government of Aragon, World Bank and IAMZ-CIHEAM.
2   The current world population is around 6600 million and income per capita is US$8000, while predictions for 2050 are that there will be 9000 million inhabitants and income per capita will be US$20,000.
3   Water is a common pool resource or impure public good. A pure public good is characterized by non-exclusion and non-rivalry.
4   A technological exclusion can appear when the costs of deepening wells become too high. This is the cost exclusion mechanism that prevented the widespread development of aquifer exploitation before the fall in pumping technology costs during the second half of the 20th century.
5   See the arguments in Chapter 4, explaining that water pricing is useless for allocating irrigation water, and that CAP does not influence high profit crops, such as fruits and vegetables, that cause the most severe water resources degradation.
6   In the EU, the strategic behaviour of countries consists in maximizing financial returns and minimizing contributions, and the same principle applies to environmental policies. The UK and The Netherlands are quite skilled at defending their interests and in climate change they are the only EU countries using a land use, land use change and forestry (LULUF) clause to increase their emissions threshold in 1990, in order to reduce their subsequent effort. The lack of information favours the adoption of strategic behaviour to avoid complying with rules and regulations, but also favours the ability of forming country coalitions to modify regulations. This was the case with the Stability Agreement infringement by Germany and France. Germany and France changed the rules because they were unable to comply with the Stability Agreement and they did not want to pay the resulting very large fines.

References

Freeman, A. (1979) *The Benefits of Environmental Improvement: Theory and Practice*, Johns Hopkins Press, Baltimore
Freeman, A. (1990) 'Water pollution policy', in P. Portney (ed) *Public Policies for Environmental Protection*, Resources for the Future, Washington, DC
Loomis, J. (1997) 'Use of non-market valuation studies in water resource management assessments', *Water Resources Update*, vol 109, pp5–9

Part I

Non-point Source Pollution Regulation Approaches

2

Problems in Assessing Non-point Source Pollution in China: Links to Policy and Regulation

Edwin Ongley and Yu Tao

Introduction

The concept of non-point source (NPS) pollution was first put forward and studied in North America and was the subject of intensive research, especially from the mid-1960s to the 1980s due, in part, to the eutrophication 'crisis' in the Great Lakes of North America. In the 1970s, although point source pollution was reasonably well controlled, the water quality compliance in surface waters was 65 per cent for rivers, 78 per cent for seas and only 42 per cent for lakes (Liang et al, 2004). This convinced researchers that some source other than the point sources was still damaging water quality. Consequently, NPS pollution refers to polluting sources that are diffuse, that is, not discharged from an identifiable point (e.g. through a pipe). This includes agriculture, street run-off, deposition of atmospheric pollutants, mine sites, transportation corridors, etc. In the US, *non-point source* is defined as meaning any source of water pollution that does not meet the legal definition of *point source* in the Clean Water Act. NPS types, amounts and best management practices became a subject of intensive research. In terms of pollution impacts, two types of NPSs are of particular concern: agricultural NPSs and urban NPSs, although NPSs can include many other types of land uses (Ongley, 1996). To this day, agricultural NPS pollution continues to be a major concern for water quality in the US (USEPA 2003a).

As a result of severe pollution in several large Chinese lakes, and a growing concern in China over agricultural sources of water pollution, research into NPS pollution started in the early 1980s with the study of urban run-off pollution in Beijing and in some polluted lake environments such as Dianchi and Taihu (Zheng and Wang, 2002). This and subsequent research has never been consolidated into systematic appraisals of NPS impacts in China. In 2004, the Asian Development Bank (ADB, 2004) released its final report of rural NPS pollution in China. That report notes the many difficulties in making a comprehensive assessment.

In recent years, and especially since regulation of animal feedlots came into force in China in 2001, there is a growing body of opinion that NPS pollution, especially agricultural NPS pollution, is a particularly serious problem in Chinese management of water quality. There is consensus in the literature that the greatest agricultural impact on water quality in China has been intensive animal rearing (cattle, dairy cows, swine, egg and meat chickens). At issue is the role and definition of 'intensive' animal rearing. The definition of 'intensive' livestock operations is notionally based on the criterion that the number of animals exceeds the number for which the waste can be effectively utilized on the farm or in surrounding areas, and is similar to the criterion used by the United States Environmental Protection Agency (USEPA) for 'concentrated animal feeding operations' (CAFO). According to a study of the environmental effects of the beef industry in China carried out in the Zhongyuan (or Central Plains) beef belt, particularly in Henan, Shandong and Hebei provinces (Liu, 2000), involving 50 households and 30 feedlots with less than 50 head of cattle, virtually all manure is utilized on the land and is not considered an environmental hazard. The ADB states that the number of animal units (where units of various types of animals are calculated in comparison with waste produced by swine) in operations of less than 50 animals comprises 76 per cent of all animal units in China and leads to its conclusion that it is the approximately 1 million intensive livestock operations that are the main concern for water quality in China. Using the State Ministry of Environmental Protection (MEP – formerly SEPA) defini-

Table 2.1 *Classification of 'intensive' feedlots in China subject to regulation*

	Pigs (> 25kg)	Chickens		Cattle	
		Laying chickens	Meat chickens	Adult dairy cattle	Other cattle
China Class I (large)	≥3000	≥100,000	≥200,000	≥200	≥400
China Class II (small[c])	500–3000	15,000–100,000	30,000–200,000	100–200	200–400
US (large CAFO)	≥2500	>30,000[a] >125,000[b]	>30,000[a] >125,000[b]	>700	>1000
US (medium CAFO)	750–2499	9000–29,999[a] 25,000–81,999[b]	9000–30,000[a] 37,500–124,999	200–700	300–1000

Note: a With liquid manure handling; b with other types of manure handling; c a small CAFO is one that has fewer animals that those of the medium CAFO. There are also US criteria for ducks, turkeys, sheep, etc.

tion that those units having more than 200 head of swine are considered 'intensive', ADB reported that such units make up only 8 per cent of total swine production. Intensive animal rearing in China is divided into two classes in Table 2.1 and these are compared with comparable CAFO sizes from the US.

Legislation and managerial measures for agricultural source control

In 2001, the MEP regulated intensive feedlots and moved these into the point source category (as did the US) and imposed controls on wastewater, odour, waste management systems, location, and other requirements such as permitting, pollution fees, environmental impact assessments (EIAs) for new feedlots, etc. The Chinese regulations contain provisions that are parallel to those in USEPA regulations that were promulgated in 2003 (USEPA, 2003b). The current Chinese regulations are noted in Table 2.2. The 10th Five-Year Plan (2001–2005) set a target of 60 per cent of wastewater from intensive livestock operation in 'key' areas to meet standards. It is curious that the USEPA (2003b) in its Final Rule states that the regulation will bring under control 60 per cent of

Table 2.2 *Regulations for feedlot and animal husbandry in China*

Regulation	Issue/implementation date	Description
SEPA Decree No 9: Management Methods for the Pollution from Livestock and Poultry Breeding	Issue date: 8 May 2001 Implementation: 8 May 2001	Focuses on principles of recycling, reuse and reduction of animal wastes, requirements for EIA, etc., role of local governments. Establishes basis for EIA, permitting, pollution fees and fines, and prohibits intensive livestock production in specified areas
Official code: HJ/T81-2001: Technical Standard of Preventing Pollution for Livestock and Poultry Breeding	Issue date: 19 December 2001 Implementation: 1 April 2002	Prescribes basic technical standards including: location, interior layout, waste cleaning techniques, waste storage and treatment, manure reuse, feeds and feeding management, disposal of dead animals and monitoring
Official code: GB18596-2001 Discharge Standard of Pollutants for Livestock and Poultry Breeding	Issue date: 28 December 2001 Implementation date: 1 January 2003 for Class I & II feedlots in prescribed areas. For other areas, local EPBs can set the implementation date but no later than 1 July 2004	Sets: daily maximum allowable concentration of water and odour pollutants; maximum allowable discharge volume of wastewater, and water quality standards
a. Livestock & Poultry b. Rural Living c. Agricultural Fertilizer and Pesticide Usage	To be issued in mid-2008	Content not yet known

Note: EPBs = Environmental Protection Bureaux

wastes generated by intensive animal operations. However, direct (ADB, 2004) and anecdotal evidence suggest that enforcement of these regulations is not very effective in China and that intensive animal rearing continues to be a serious problem for water quality.

Trends in NPS research

The first comprehensive analysis of measures to control agricultural pollution was published by the United States Department of Agriculture in 1976 (USDA, 1976) and included a statement of additional research needs. In part, this was built onto programmes established during the United Nations Educational, Scientific and Cultural Organization's (UNESCO's) Hydrological Decade (1965–1974) during which North American researchers established many hundreds of 'experimental' and 'representative' river catchments at differing scales to better understand hydrological processes. Since the late 1970s there has been extensive research, especially in North America, into the process dynamics of various types of NPSs, most notably those of agricultural and urban run-off. This research parallels major advances in erosion and sediment run-off dynamics which produced the now famous Universal Soil Loss Equation (Wischmeier, 1976) – which remains today as a central component of many NPS models – and the 'sediment delivery ratio' (USDA, 1983), which remains a critical concept within NPS models, especially Export Coefficient Models.

Since the 1980s and as a direct result of the many empirical studies of sediment and chemical run-off processes in agricultural systems at different scales (plot, field, catchment) generated across North America, there were major developments in NPS dynamic modelling. Some of the better known dynamic models are noted in Table 2.3. Except for MIKE-SHE, these models are public domain models widely used in the US and internationally, including China.

Table 2.3 *Summary of better-known agricultural NPS models*

Model	Name	Reference
AGNPS	Agricultural non-point source	Young et al (1994)
ANSWERS	Areal Nonpoint Source Watershed Environment Response Simulation	Bouraoui and Dillaha (1996, 2000)
BASINS	Better Assessment Science Integrating Point and Non-Point Sources	USEPA (1996) (with updates)
CREAMS	Chemicals, Run-off and Erosion from Agricultural Management Systems	USDA (Foster et al, 1980)
MIKE-SHE	Proprietary model developed by Delft Hydraulics	www.dhisoftware.com/mikeshe
PLOAD	Pollutant Loading Application (GIS screening tool)	Users Manual, V3, USEPA (2001)
SWAT	Soil and Water Assessment Tool	USDA (Neitsch et al, 2002)

In the past decade, and in contrast with empirically based dynamic models, there has been parallel development in other types of models, especially the Export Coefficient Model. This approach is a variation of the Unit Load approach that was initially developed in the US in the early 1970s before there were extensive empirical data on types of NPSs and rainfall–run-off relationships (Uttormark et al, 1974). Export coefficient modelling is a river basin-scale, semi-distributed approach which calculates mean annual total nitrogen and phosphorus loading delivered to a water body. This method calculates the sum of the nutrient loads exported from each nutrient source in the river basin (AERC, undated). The model (Johnes, 1996) allows empirical plot data to be scaled up to catchment scale.

The Export Coefficient approach is gaining favour because of its simplicity and relative robustness; the time step is large (months, seasonal or yearly – allowing use of spatially and time-based lumped data rather than real-time data, and agricultural census data rather than field level data). Note, however, that the model offers improvement over classic unit load models only when there are enough empirical data at the plot scale to lead to empirically derived export coefficients (E) for various combinations of crop type, precipitation, land management, soil type and topography. In this regard, the empirical requirements are comparable to those of dynamic/process models. The Export Coefficient approach is being used at catchment and at national scales in the UK (AERC, undated), the US (NC-DENR, 1998; Endreny and Wood, 2003) and China (Ding et al, 2003; Ma et al, 2005). Export coefficients are an essential part of the geographic information system (GIS)-based NPS screening model PLOAD (USEPA, 2001).

In recent years, Chinese researchers have been focusing on NPS pollution assessment techniques using various empirical and statistical models such as GIS NPS modelling (Cui et al, 2003), SWAT and related modelling approaches (Cui et al, 2003; Hao et al, 2006) and export coefficients. There is a growing body of empirically based studies of water quality relative to different NPS pollution land uses (Cui et al, 2003; Shuai and Xia, 2006; Duan et al, 2006). Nevertheless, the amount of comprehensive empirical investigation of NPSs in China remains small relative to that in North America and has not been consolidated into regional or national databases (as, for example, US data under various USEPA and USDA research programmes). In comparison, as early as 1979 the USEPA was able to calculate the contribution of NPS pollution to eutrophication using 928 sampling sites that had been established nationwide in small catchments having no point sources (USEPA, 1979).

NPS loadings

One important aspect of NPS pollution study is to quantify the load. In view of its characteristics, however, the estimation of NPS pollution load is far more difficult than point source load. In particular, as the scale increases, estimation of total NPS contribution to pollution loads becomes increasingly difficult (Ongley, 1987). Empirically based simulation modelling of NPS pollution goes back at least to 1976 when USEPA published its Non-point Source Pollutants Loading Model (Donigian and Crawford, 1976). Since then many models have been developed in North America based on a vast amount of empirical data at plot, field and drainage basin levels throughout the US and Canada (Table 2.3).

Many of these techniques have been used to a greater or lesser degree worldwide. In China these have been used, but often without the very extensive empirical database required to calibrate the techniques properly. Indeed, we conclude that in parts of China, such as the north-east and North China Plain areas, the physical conditions (land surface and hydrological regimes) and agricultural typologies are so different from those in North America that the application of these techniques is likely to produce quite erroneous results. We explore this further below.

In North America, where point source pollution is well controlled, there is no dispute over the central role of NPS pollution in water quality; however, the estimates of NPS pollution loadings vary greatly. Nitrogen load from NPSs was estimated at 33–63 per cent of the total nitrogen load, and phosphorus load from NPSs was 42–59 per cent of the total phosphorus load; these numbers include nitrogen and phosphorus loss caused by physical and geochemical processes (Smith et al, 1997). In a study of the Greater Mississippi Basin, Goolsby et al (1999) reported that 89 per cent of the nitrogen loading from the Greater Mississippi River basin to the Gulf of Mexico can be attributed to non-point sources including fertilizer, erosion, groundwater discharge, animal waste and atmospheric deposition. The remainder comes from municipal and industrial point sources. Phosphorus inputs to the Gulf of Mexico come from fertilizers (31 per cent), animal waste (18 per cent), point sources (10 per cent), and generalized basin run-off (including soil erosion) (41 per cent). In their 12-year study of water quality in the Upper Mississippi, Garland et al (1999) found that NPS pollution contributed from 66 per cent of total phosphorus in high flow years, to 20 per cent in low flow years.

In Chinese studies, the non-point source load estimates range from 35–55 per cent of the total load, to 65–75 per cent or more (Liu, 2000; Wang et al, 2002; Y. Chen et al, 2003; Jin et al, 2004; Cai et al, 2005; Cheng et al, 2005, 2006b). ADB (2004) found that rural NPS pollution in China contributed mainly chemical oxygen demand (COD), total nitrogen (TN) and total phosphorus (TP), with rural NPS COD levels in 2001 about 1.42 times all urban and industrial point sources of COD. However, the ADB methodology is based mainly on available data, does not adequately define what is included as rural NPS pollution, and presents NPS pollution estimates that we believe to be flawed for the many reasons noted in this paper. Chinese scientists, using SWAT Model outputs, presenting data at a working meeting on NPS pollution in a large basin on the North China Plain,[1] proposed that NPS contribution to total nitrogen was 52 per cent and to total phosphorus was 76 per cent. The classes of land uses included as NPS were not known; however, the values were defended on the simplistic assumption that some 75 per cent of the basin population is 'rural'.

Hao et al (2006) have developed mathematical representations of the dynamic components (run-off, etc.) for estimating components of NPS pollution for major river basins in China, using a hybrid approach in which the various coefficients are from SWAT, foreign literature and some limited investigations. They report that the NPS contributions range from 58–73 per cent for TN, 66–93 per cent for TP and 67–86 per cent for ammonia (NH_3-N). Of these NPS components, farmland is thought to be the source for 73–86 per cent of TN, 64–91 per cent of TP and 70–86 per cent of ammonia. In perhaps the most recent study of NPS pollution in the Three Gorges Reservoir for 2004–2005, Chinese scientists have reported that NPS contributions to total load are $\geqslant 66$ per cent for TN, $\geqslant 90$ per cent for TP and, incredibly, 87–90.3 per cent for COD.[2] It is true that there will be a wide range of reported values for NPS pollution depending on the magnitude of point sources within individual catchments; however, our main point is that the overall range of values and especially the upper ranges presented, are much too high to be reasonable, especially for larger river basins.

In addition to problems of the methodologies used in many Chinese studies, the high proportion of the NPS contribution to total pollutant load that is commonly reported in China seems unreasonable when, unlike the US, point sources remain poorly controlled. We believe, therefore, that it is unreasonable that NPS, and especially farmland, could be such a large percentage of total pollutant load. An NPS of 87–90.3 per cent of COD in the Three Gorges Reservoir is simply not credible unless it includes untreated urban wastewater that is routed to adjacent watercourses via many drains. In western studies, this would generally be classified as a point source; however, it is not clear in Chinese studies if it is classified as a point source or an NPS. We conclude that uncritical use of the NPS category in many published studies, without defining what it includes, can lead to illogical and perhaps indefensible statements on the role of NPS in Chinese water pollution. The issue is significant in that NPS pollution estimates in China are beginning to impact on public policy development regarding the best strategies for pollution control. For this reason, it is imperative that NPS pollution estimations be based firmly on credible technical procedures and have clear definitions of what types of land uses are included. If, for example, untreated urban wastewater is being classified as an NPS and contributes most of the COD, then the policy result should be focused on urban wastewater treatment and not on rural NPSs.

As a result of these discussions, we began investigating in more detail the nature of Chinese NPS studies, the types of empirical data used, and the physical situation in China relative to that of North America, with the objective of determining inherent difficulties and limitations in the use of North American NPS estimation techniques in China. The analysis began with a review of a large number of NPS papers published since 1994 in Chinese journals and some unpublished research reports. Based on our knowledge of North American NPS estimation techniques, we report here our observations of problems in Chinese NPS methods, and focus on the different conditions for applying load estimation methods in North America and China and the probable errors resulting from using these methods in China. The paper also proposes recommendations for dealing with these problems.

Problems in non-point source studies in China

There are five commonly used estimation techniques that have been applied in China.[3] These are:

1   total measured load in a section minus reported point source loads;
2   Unit Load Method (including Export Coefficient modelling);
3   Hydrograph Separation Method;
4   SWAT Model estimation;
5   mathematical representation.

These five methods, or their modifications or in combination, cover almost all NPS load estimation methods presently used in China and are summarized in Table 2.4.

Total measured load in a river section minus reported point source loads

This method is simple in principle but more difficult in practice. One of the problems in this method is common to most NPS studies in China – namely, the problem of defining 'point source'. This has a substantial impact on the final result. When this method is used, it is usually not clear whether non-sewered small cities are included as NPSs or not. In North America, virtually all urban areas, large or small, use some form of centralized wastewater treatment and are classified as point sources. Only very small villages and rural areas where householders use individual septic systems and ground disposal of human wastes are classified as NPSs. Urban NPSs in North America are mainly restricted to urban rainfall (street) run-off and have been the subject of major research efforts (usually in the context of 'combined sewer overflows', CSOs) due to the toxicity observed in urban NPS run-off. In China, the distinction between urban point and NPS is not simple in that many cities have partial or no sewage treatment and discharge human and industrial waste into adjacent rivers or lakes via a

Table 2.4 *Summary of NPS estimation techniques commonly used in China*

Method	Description	Advantages	Disadvantages	Reliability[a]
1 Measured total load, less reported point loads	Uses measured load in river cross-sections (total load) minus reported point source loads = NPS load	Data are available for river discharges and pollutant concentrations	Reported point source loads are often low, sometimes very much less than actual. May or may not include large areas of unsewered urban black water	Low (errors known to be up to 45%)
2 Unit Load and Export Coefficient Method	Pollutant load generated per unit of cow, pig, human, etc. Now used with export coefficient to allow for hydrologic transport from source to stream	Easy to use. Unit loads well known. Can be scaled up to large watersheds	Export coefficient not usually known in Chinese situation, especially as agricultural land management practices are so different than in US or UK situations	Medium to low
3 Hydrograph Separation	Uses conventional hydrograph separation technique	Technique is well known to hydrologists	Storm flow/base flow separation does not discriminate between NPS (storm flow) and point source (base flow) in Chinese situation	Low
4 Modelling	Many types of empirically based models available that link various land use types to water, sediment and pollutant run-off at field and catchment level	Models are well known and documented	Models based mainly on US agricultural conditions and land use practices which are very different from those in China	Medium
5 Mathematical representation	These are developed by various Chinese researchers and attempt to simplify complex relationships into mathematical equations	Use many simplifying assumptions often based on the user's academic understanding of the linkages between water, sediment, pollutants, etc.	Generally, these are not adequately explained or documented in Chinese use. Simplifying assumptions probably do not reflect the field level situation	Not known, but probably low

Note: a Authors' estimate of reliability under current use in China.

multitude of street-ditches, pipes and sewage canals. In China, there is no accepted definition of whether this type of urban run-off should be classified as point or non-point source. We suggest that all of these types of urban discharge should be classified as point sources as they are controllable sources in comparison with, for example, agricultural NPSs which are mainly non-controllable, diffuse sources. However, as this situation is unique to China (and many developing countries), Chinese scientists and policy makers should try to reach a consensus on the definition of various NPSs. The lack of a common definition of 'urban' point source can lead to incorrect assumptions about the role of what is technically accepted as NPS, and consequently exaggerate the NPS load. For example, this method was recently used in Songhua River Basin and the estimated NPS load was about 45 per cent of the total load (ADB, 2005), but we believe this result may be an overestimation for the reasons noted.

A second problem is that point source waste loads reported by Environmental Protection Bureaux are often under-reported. This can be due to a variety of reasons such as under- or fraudulent reporting by industrial enterprises, lack of loadings information for township and village enterprises, illegal discharges, etc. In the Huai River, for example, under-reporting of actual loads is estimated to be up to 45 per cent in comparison with waste loads measured by the Ministry of Water Resources (MWR).[4] For 2005, official statistics of the MEP and the MWR indicate that total wastewater discharge reported by the MEP for China was 52.4 billion tons compared with 71.7 billion tons reported by the MWR – an under-reporting by the MEP of 36.8 per cent (see Chapter 7) insofar as MWR data are considered by many experts to be more accurate.

Unit Load and Export Coefficient Methods

The Unit Load Method calculates the load in accordance with known values of the chemical content of excreta by humans, domestic animals and fowl (chickens, ducks, etc.). These values are multiplied by the population of humans, animals or fowl to estimate the total *potential* NPS load. As noted above, this method was developed in the early 1970s in the US (Uttormark et al, 1974). The challenge in this approach is the technical requirement that there must be a hydraulic connection between the source and the aquatic environment, that is, rainfall or spring snowmelt produces surface run-off that carries all or some of the load to adjacent watercourses. In the Unit Load approach, a 'delivery ratio' or 'export coefficient' must be applied to account for the degree of hydraulic connection. Load calculated this way becomes NPS load only when the source is, in fact, carried to a watercourse. If there is no run-off, there is no NPS load irrespective of the size of the human, animal or fowl population. This method, in its simplistic form, is not used much now because of the very large variance in results produced by such uncertainties as geomorphology, hydrology, run-off, scale effects, basin size, etc. In 1976, the USEPA published a much more comprehensive approach to loadings functions which is based on extensive empirical data and requires the user to input site-specific characteristics in order to determine delivery ratios and, in turn, the load that is delivered to the watercourse (McElroy et al, 1976).

Over the years, the Unit Load methodology has evolved into the Export Coefficient Method – and related models such as AERC (undated) in the UK, and PLOAD (USEPA, 2001) in the US. The export coefficient is determined by linking unit loads of particular sources with export coefficients that have been empirically observed, mainly in North America and more recently in the UK. GIS-based models such as PLOAD (USEPA, 2001) contain export coefficients based on empirical research over many years in America.

The problems in the use of this method in China are mainly associated with the problem of determining what amount of the source is transmitted to the watercourse (hydrological connectivity). This is usually unknown and difficult to estimate. In the north, north-east and west of China, water scarcity has led to an agricultural landscape that is designed to retain water and there is virtually no run-off except under exceptional storm events. In the humid south, extensive terracing and paddy rice culture also conserves water. Use of some arbitrary proportion of total run-off as the run-off factor is not a satisfactory solution in that the timing of the run-off in relation to rural land use is at least as important as the actual run-off amount.

It is not well known to Chinese researchers that, in much of agricultural North America, the problem is excessive water and large parts of the agricultural landscape are designed to get rid of water through tile drains installed under the fields and in ditches adjacent to fields to prevent waterlogging. In water-scarce parts of China, ditches are for irrigation water supply, not for drainage, and the fields are surrounded by low berms to retain irrigation water and prevent water run-off. Furthermore, in North America the field size is very large, allowing run-off patterns to emerge after rainfall. In comparison, in China the fields are very small (e.g. family plots) so that run-off cannot easily develop. For example, our observations of field geomorphological evidence from the area around Yuqiao Reservoir near Tianjin (an area of rolling topographical relief), indicate that there had been little or no run-off into the reservoir from surrounding fields, at least in the past year. Furthermore, the entire agricultural landscape is designed to retain water and to prevent run-off, with berms placed across small valleys that, many years ago, carried run-off. There are no channels or other evidence of confined flow or surface flow in these valleys and no evidence of overland flow on adjacent fields or run-off from fields (as would be evidenced from erosion features). In this case, therefore, we conclude that there has been little or no NPS load at least within the past year and the export coefficient for surface transported pollution would be zero. We do not know if there has been leaching and subsur-face or groundwater transfer of pollutants to nearby waterways. In southern humid China, the extensive areas of small terraced paddy rice ponds and other forms of terracing on arable slopes are quite different from the arable landscape of North America. Paddies are designed to retain water and are drained only to harvest the rice. Export coefficients contained in models such as PLOAD are not directly applicable. The work of Cui et al (2003) in an agricultural catchment in

Taihu Lake basin is a good example of empirical research in China that 'works around' the limitations of PLOAD by providing empirical values for EMCs (event mean concentrations of various pollutants) from their field site.

Hydrograph Separation Method

This technique relies on the 'run-off effect' in which, in North American literature, the storm flow component is associated mainly with non-point sources and base flow is mainly associated with point sources. Therefore, in principle, separation of a run-off hydrograph into base flow and storm flow (component in excess of base flow) should permit separation of the point source and NPS loads. In North America, this method is usually restricted to small basins where there is no surface storage (dams and reservoirs) and there are no other data with which to make the NPS estimation. In such situations in North America this technique works adequately as an estimation technique. Storage systems complicate the result because water released from storage systems during storm events contains an accumulation of pollutants from point sources.

The problems with this methodology in China are mainly:

- Most Chinese rivers have barriers or reservoirs, behind which point source pollutants accumulate. Water management in China requires that reservoirs (that become grossly polluted during the dry season) be flushed downstream during or in advance of high flow events. Even where there are no major reservoirs, smaller Chinese rivers contain large quantities of point source pollutants that are not moved downstream in significant quantities until there is a flood event. This also includes mobilization and transport of highly polluted bottom sediments during storm events. Therefore, downstream, these appear as the 'run-off effect' cited in western literature, but in fact are mainly from the release of accumulated point sources from storage areas during rainfall–run-off events. Therefore, hydrograph separation is not a reliable method to identify NPS loads in China.

- When applied to large rivers such as the Yellow River, similar problems emerge in so far as the storm flow component contains not only NPS pollution but also accumulated point source pollutants, especially in bottom sediments, from tributary rivers that are mobilized during storm flow periods. In fact, some tributaries have become mainly wastewater rivers due to heavy pollution, as for example, the Weihe (river) and Fenhe (river) that are tributaries to the Yellow River in Shaanxi and Shanxi provinces, respectively. Also, large rivers such as the Yellow River have multiple reservoirs that contain point source pollution that is released during high flow events. Under these circumstances, this methodology cannot accurately separate point and non-point sources.

- A further major difference between North America and China is that rivers in North America are perennial with sustained base flow and low levels of pollution. This makes the base flow an easy estimator of point sources. In China, however, many rivers are seasonal or have zero flow in dry seasons, especially in north and north-eastern China. In such rivers where there is base flow, it comprises mainly wastewater and is not true base flow (which is sustained by groundwater). The water chemistry is diluted during storm events so that the point source component also appears in the 'run-off component' and not just the base flow component.

We conclude that use of the Hydrograph Separation Method is likely to greatly overestimate the NPS component. The magnitude of the error is not known in the Chinese context but we believe that it would be quite large. An example is the estimates provided for NPS contribution to the Three Gorges Reservoir (noted above).

The use of the three methods discussed above exaggerates NPS loads in many areas of China. Where we can compare these models within the same study, the three methods seem to produce roughly similar values of NPS load in that study, which, we assume, probably reflects similar but incorrect scientific assumptions.

SWAT as an example of NPS estimation using dynamic modelling

While a variety of process models have been used in China, SWAT demonstrates a common set of problems with model application in the Chinese context. In North America, the wealth of empirical research into rainfall–run-off relationships, erosion and sediment transport, and run-off–chemical transport relationships at plot, field and basin levels, has led to the development of many models for estimating NPS load (Cheng et al, 2006a). Among these models, the SWAT Model is widely used in North America and elsewhere, including China (Hao et al, 2006). It is a distributed, physically based, watershed-scale model, incorporating considerations of the climate, surface and underground run-off, soil type, vegetation growth and agriculture management in the modelling of the NPS load. Below, we focus on the unique problems in the use of the SWAT Model in China and, in particular, how differences in conditions between North America and China can compromise the results in the Chinese context.

Empirical knowledge base required for SWAT

The modelling framework for NPS studies has been under development for more than 40 years in North America. The primary module in physically based models, such as SWAT, is a rainfall–run-off module that routes rainfall across the field surface to adjacent watercourses. This component in SWAT takes digital elevation data, together with land use, soil information, crop types, agricultural management techniques, etc. and stores these relationships within the model in 'look-up' tables. The user of the model inputs the spatial characteristics of his study area into the model and the coefficients required by the model are derived automatically by the model from internal look-up tables. The model then uses these values to calculate run-off (and erosion and chemical transport). This process works well in the US because the module has been calibrated and validated in thousands of situations (different land uses, different crops and vegetations, different precipitations and different agricultural managements, etc.) and in different hydrological response units (HRUs). This vast amount of empirical information makes the parameters in the look-up tables adaptive to the calculation of run-off under most situations in the US. As a result, the application of the model in the US can derive a reasonable value for run-off at any HRU for any given quantity of rainfall. In China, there is little empirical data with which to validate the look-up tables, so the Chinese user is confined to physical characteristics within the look-up tables that reflect the US landscape.

Hydrological connectivity

The application of the SWAT Model assumes a more or less continuous process from rainfall to run-off, that is, the hydrological connectivity is continuous from plot, to field, to small catchment, to large catchment for any rainfall that produces run-off (rainfall intensity exceeds infiltration capacity). There is the provision in SWAT for interruption of run-off by reservoirs or ponds (Neitsch et al, 2002), but it is not designed for the lack of flow continuity that occurs, for example, throughout much of Chinese agricultural land where the land surface is designed to retain, not transmit, water. As noted below, run-off calibration at the basin scale will ensure that the aggregate of flow from all upstream land surfaces will be correct, but there remains great uncertainty about the run-off from individual land units which produces major problems for the estimation of chemical and sediment run-off. In our review of Chinese work, researchers have paid little attention to these issues.

Calibration and validation

Run-off calibration in the Chinese application of SWAT is only at the basin scale using hydrometric data from river gauging sites. These gauging sites are substantial distances downstream from headwater areas and are rarely indicative of any one type of land use or HRU. While the model calibration will force the model to produce correct results at this large scale, what is not known and cannot be determined is whether the different HRUs and different types of land surfaces upstream of the gauging station are contributing the correct run-off values for each land type. The calibration used in Chinese applications only ensures that the aggregation of land types produce the correct run-off at the scale of the basin area at the gauging station. While this is not important for run-off estimations at the basin scale, it becomes very important for pollutant modules of the model in which chemical contribution depends very much on land and land use characteristics at a much smaller scale which can capture the detailed relationship of land, land use and land management that is required by the model and contained in the look-up tables. In fact, few of the Chinese crop and crop management types are represented in the run-off curves (coefficients) contained in SWAT.

In Canadian studies, SWAT has required extensive recalibration for NPS pollution estimations in areas for which the look-up tables are not appropriate. In China, the SWAT Model application is used without the extensive empirical evidence required to recalibrate the model and to modify the look-up tables. In light of the very different land-use management characteristics in large parts of China in comparison with North America as noted above, and in the absence of the extensive research required to correct the look-up tables to the Chinese situation, it seems unlikely that SWAT will provide particularly good estimates of rural NPS pollution loadings except under conditions that can be demonstrated to be similar to North American conditions. This, therefore, makes it imperative that users of the SWAT Model in China ensure that their application area is sufficiently similar to US conditions in order for the model to be expected to provide reasonable NPS estimates. From our observations, the main problem will be model calibration in agricultural areas which is also the main source of rural NPS pollution.

Artificially managed flow
The application of dynamic models on flat areas such as the North China Plain is difficult in so far as there is little to no 'natural' hydrology. All rivers are canalized with flow routed artificially between control structures (sluices). For much of the year, these rivers may be dry as there is little to no base flow due to greatly depressed groundwater tables; where flow exists it is mainly wastewater from upstream point sources.

Point sources
It is not clear to us how SWAT users have incorporated and calibrated point discharges within SWAT although there is provision for this in the SWAT documentation (Neitsch et al, 2005).

Irrigation
Although SWAT does deal with irrigation, the input requirements are substantial. For small irrigated family plots that are typical of the North China Plain, the ability to input the required data may be impossible except perhaps on a 'representative basis' with extrapolation to similar areas. Large irrigation districts such as Qingtongxia and Hetao in the upper and middle reaches of the Yellow River, for example, are known to have substantial impacts on water quality of the Yellow River (J. Chen et al, 2003) and may be more amenable to the use of SWAT.

Definitions
SWAT application also suffers from lack of definition (discussed above) of what constitutes point and non-point sources in the Chinese context, and how

agricultural and other rural land uses are differentiated in deriving NPS results.

The discussion above does not mean to deny the applicability of the SWAT or similar models in China, but to emphasize that full attention must be directed to the differences between American and Chinese conditions. We believe that a large amount of empirical evidence will be required to adequately apply SWAT for pollution identification and management.

Mathematical representation

These are, usually, static equations using coefficients to represent run-off, contributions from different types of sources, etc. These are integrated into equations with the objective of estimating loads from NPSs. This approach is used to estimate different components of non-point source loads for each of the major river basins in China (Hao et al, 2006). This method suffers from the fact that it is static (does not allow for time-dependent changes in run-off, for example). The constants and mathematical relationships are estimated using SWAT, or from published foreign literature, and/or from limited field investigations. However, given the lack of empirical data from different types of NPSs in China, the results cannot be said to be calibrated or verified. The results are probably not much more reliable than the Unit Load methodology using run-off (export) coefficients which, for most agricultural surfaces in China, cannot account for the difference in run-off behaviour in comparison with the US from which run-off coefficients (as in SWAT) are derived.

The many inherent problems in using the five methods described above suggest a variety of key issues that need to be addressed by Chinese researchers before the role of agriculture in water pollution and, more generally, the role of NPS pollution, can be estimated with some degree of certainty. These are explored below.

Considerations for agricultural non-point source studies in China

Domestic and foreign studies show that agriculture is a major source of the NPS load (Agrawal et al, 1999; Zhang et al, 2004). In fact, there is no simple way to estimate NPS load from agriculture, e.g. fertilizer loss with run-off. In North America this has been the subject of extensive research, leading to a variety of models, but these models must be calibrated for each site, then verified before use. Because little of the necessary data are available in China to easily apply such methods, we make the following suggestions for agricultural NPS studies and which should be considered in making estimations of NPS loads from agriculture.

Rainfall–run-off mechanisms

The conditions under which Chinese agricultural surfaces produce run-off need urgent empirical investigation. This is a critical issue that must be resolved in order to determine sediment, nutrient and pesticide run-off from agriculture. It is possible that, in contrast to North America where agricultural surfaces provide a large proportion of the run-off, in China the main source of run-off from rural areas may be from non-agricultural surfaces (mountains and non-arable hilly terrain) except in situations of very large rainfall events. Western hydrological models (such as that in SWAT) need to be reformulated to capture the particular 'water conservation' characteristics both of water-scarce areas and the unique conditions of terracing and paddy fields in humid areas. Similarly, expert coefficients developed in western countries will not directly apply to China. These will require much field investigation and measurement as, we believe, it will be found that these types of agricultural surfaces will retain all rainfall up to a certain threshold which, when exceeded, will then produce significant run-off. This is a step-function that can only be defined from empirical research. A critically important question is how often is the critical threshold that produces run-off, exceeded. Knowledge of this will allow much more accurate modelling of run-off and, in turn, of pollutant transport from agricultural surfaces.

Irrigation run-off

It is known that large irrigation districts have measurable impact on receiving water bodies (e.g. Yellow River; J. Chen et al, 2003). The impact of this type of agricultural pollution source both as overland flow and as subsurface run-off needs much more investigation, including the North China Plain where fields are small and there is little measurable overland flow.

Empirical studies

There is an urgent need for a wide range of empirical studies of sediment erosion and transport, and sediment-associated chemical transport in all significant agricultural areas of China. There are critical differences between North America and China in this regard. Sediment erosion is non-existent if there is, in fact, no run-off from water-conserving agricultural areas. Western erosion models assume that any rainfall, once it exceeds infiltration capacity, will produce run-off as fields are designed to eliminate water. This is not true of many agricultural landscapes in China as most run-off is captured within the field boundaries. Although Chinese farmers now use large amounts of agricultural chemicals, these are not eligible as an NPS load if there is no run-off. Therefore, these relationships need to be established in order to meaningfully apply standard erosion and chemical transport models.

Loss of nitrogen is a substantial agricultural NPS problem in North America and much of this is through subsurface tile drains. In China, tile drains are not normally used, therefore the nitrogen run-off component needs to be established and related to nitrogen fertilizer usage at the field scale. There will be substantial regional differences in China in nitrogen run-off, especially comparing drainage of former wetlands, now farmland, with shallow groundwater in, for example, Heilongjiang, relative to deep groundwater areas of the East China Plain. Pesticides in North America are applied via large machinery on fields that are often several square kilometres in area. In contrast, in China, pesticides are applied mainly by hand on small plots. It is not known if these contrasting management practices have an impact on how pesticides contribute to NPS pollution in China. Also, in view of the water-conserving nature of many agricultural landscapes, modern pesticides can only contribute to NPS pollution if they are applied in the few days before the run-off threshold is exceeded.

Chemical run-off

Although a large amount of fertilizer and pesticide is used in China, it is necessary to identify, in various areas, the potential for run-off and the percentage of the applied chemicals contained in run-off. In north China, some factors are particularly important:

- Rain-fed agriculture on the North China Plain is designed to conserve water so that normal rainfall produces little run-off in these areas.
- Irrigation agriculture throughout the North China Plain produces no surface run-off due to poor water availability and high cost of pumping.
- In humid areas the extensive terracing of agricultural land and paddies also produces little surface run-off under normal rainfall conditions.

Therefore, the linkages between fertilizer and pesticide application to run-off is not direct as it is in North America and NPS models must be adapted to these characteristics that are unique to Chinese (and other south-eastern Asian countries) land surfaces. The extent of leaching and subsurface run-off in humid (or seasonally wet) areas of China requires much more research in order to quantify leaching losses.

Nitrogen fertilizers are soluble and will be mobilized downwards into the soil. However, as the groundwater table is drawn down in large parts of north and north-eastern China, and as there is little or no base flow from groundwater to rivers in these areas, any enrichment of groundwater by nitrogen will have no impact on surface waters in these areas. This is very different from North America and Europe where groundwater sustains river base flow and pollution from subsurface drainage (e.g. tile drains) does affect surface waters. In humid parts of China, leaching of nitrogen into soil water and then into adjacent ditches is probably important but is difficult to quantify, and unless monitoring is done in small rural basins that have no point sources, fertilizer nitrogen is not distinguishable from nitrogen forms from other types of source. For example, the nitrogen from atmospheric deposition usually accounts for a substantial percentage in a watershed (Benjamin et al, 1991); this is likely to be especially true in China due to extensive air pollution. Atmospheric nitrogen must be separated from agricultural nitrogen, otherwise nitrogen estimates are meaningless.

Phosphate fertilizers are usually rapidly adsorbed to soil particles and do not run off except as part of the erosion–sediment transport process during rainfall–run-off events. As much of the Chinese agricultural land surface is designed to retain water, there is virtually no run-off for much of the year; in these circumstances, phosphorus run-off is likely to be almost zero. In some special areas, phosphorus loss with soil erosion is intensive as, for instance, the Loess Plateau in the Yellow River basin. However, such loss should be differentiated from the loadings caused by agriculture activities.

Modern pesticides are designed to degrade quickly. Therefore, if there is no run-off for a week or more between application and rainfall–run-off, depending on the nature of the active ingredient, there is likely to be little active ingredient that will be measurable downstream. Some banned organo-chlorine pesticides are believed to be still manufactured and, presumably, used illegally in China. These are highly persistent and will be measurable for long periods of time after application (often for years as in the case of DDT). However, these are rapidly adsorbed by fine-grained sediment and, unless there is erosion and sediment run-off from agricultural surfaces, these banned pesticides will remain at the field level, so these chemicals may not be detectable in the water environment. The application of these chemicals can be assessed only from field data. Note also that water chemistry will not reveal organo-chlorine pesticides as they are tightly bound to particulate matter; these must be analysed directly from suspended particulates taken from the water. For agriculture on upland areas (e.g. Dianshi Lake), the situation is different and fertilizer and pesticide losses through erosion and sediment transport can be significant, and eventually enter the water environment. However, this is not universal in China, and especially not the case in the North China Plain.

NPS definition

In much Chinese NPS literature, authors do not define what they include as point and non-point sources of pollution, or what is included as 'agricultural' NPS. This frustrates public policy decisions on NPS controls. For example, does 'agricultural' NPS include rural agricultural villages that rely on septic tanks and seepage pits for disposal of domestic wastewater? In most Chinese studies, there is no indication of the relative contribution of agricultural activities that are, relative to total NPS contributions, from rural dwellings. Yet, many of the published Chinese studies appear to include both agriculture and rural dwellings as 'agricultural' NPS which we believe is an uncritical use of the term and only leads to further confusion. Clearly, these two types of source are vastly different in that human habitation tends to mainly produce nutrients and pathogens in human waste, whereas agriculture is associated with sediment, nutrients and, potentially, a wide range of contaminants from agricultural chemicals.

The distinction becomes important in the context of the very large percentage of the Chinese population that lives in such villages across the country. However, we have observed in some small villages that there is relatively little direct run-off from human waste, which is in contrast with small cities and large villages in which human waste is routed to nearby watercourses. If one is interested in using NPS estimates to create public policy on pollution control options, the difference between agriculture and rural dwelling becomes quite significant in so far as, in North America, it was found that agriculture was, and remains, a major challenge for NPS control whereas villages were found to be a more effective target to reduce nutrients through various control interventions.

One major definitional problem appears to be small cities (large villages) in which most human waste is routed to nearby watercourses by drains or canals. If these were to be classified as NPS, they need to be clearly specified as a unique type of *urban* NPS so as not to confuse these with rural villages in which human waste is mainly disposed of in leaching pits, septic systems or closed lagoons. This latter type of source should be considered rural NPS, but not agricultural NPS. Agricultural NPS should be restricted to pollution from farming practices. The boundary between rural dwelling NPS and urban NPS (as described above) is not clear and one may easily be confused with the other unless field studies are carried out.

In some work, streambed sediments are referred to as a non-point source (Han et al, 2006) when, in fact, the pollutants observed in bed sediments are mainly from upstream point sources. Sediment is a storage and conveyance mechanism and not a source per se, and in our opinion, should not be categorized as a non-point source. However, as these various issues tend to be unique to the Chinese situation, Chinese researchers need to establish a common set of criteria.

Conclusion and policy implications

Non-point source and point source pollution always coexist in the environment. Experience in developed countries has shown that as point sources are brought under control, the proportion of total load from NPS becomes significantly larger *as a percentage of total load*. This is inevitable, especially as agricultural sources of NPS pollution have proven to be particularly difficult to control. Nevertheless, there are some lessons from developed counties which suggest issues that Chinese researchers and policy makers should consider.

Pollution control policy

In China, the urban wastewater treatment rate by central treatment plants is now approximately 40 per cent (ADB, 2008). Because point source control is not a technical challenge, the best approach to pollution mitigation in Chinese aquatic systems should remain focused on point sources. The environmental target in the 10th Five-Year Plan (2001–2005) was not achieved, and the goal of discharge reduction in the 11th Five-Year Plan is facing significant challenges. In the 1970s, point source control was the public policy objective in North America but with research focused on NPSs as a means of achieving greater environmental benefits, especially in locations where NPS could be demonstrated to be a major factor. Today, in the US, programmes of erosion control, withdrawal of agricultural land from use, reforestation, and best management practices in agricultural activities, are part of the USEPA's programme of reducing NPS pollution. In contrast, China remains at a relatively early stage in national point source control, therefore western experience suggests that continued focus on point source control should be the major policy position as any benefits from NPS control are likely to be small until point sources achieve at least a 75 per cent control rate. Nevertheless, there are special circumstances that warrant focused attention on NPS issues in China, especially in humid areas with greater run-off, and in lake basins where eutrophication will require both point and non-point source control to achieve an effective outcome. An example is the Lake Taihu basin in which the major cities of Suzhou and Wuxi have achieved >90 per cent urban wastewater control, in which case the role of NPSs in the continuing occurrence of severe algal blooms in the lake may, arguably, be significant.

Inter-sectoral cooperation

In North America, extensive collaboration and cooperation among departments/ministries of water, agriculture, environment and forestry, and universities have been essential in developing the knowledge required to manage agricultural non-point sources. This is a model that China should consider emulating.

Leadership of NPS work in China

Currently, there is no agency that has exercised strong leadership in NPS research and analysis in China. Many of the issues raised in this paper would normally be addressed by a lead agency in consultation with the research community. The most pressing issues, in our opinion, are:

- an agreed set of NPS definitions, creation of a NPS studies registry (who is doing what, and where), creation of a national NPS information database;
- establishing a rigorous set of evaluation criteria for NPS studies before they can be included as part of a national studies database; and

- identification and promotion of investigations into key research issues that are unique to China such as critical run-off thresholds for various types of agricultural surfaces.

Definitions

As noted above, the problem of different interpretations of what does, and what does not, constitute an NPS contributes to policy confusion.

Atmospheric nitrogen

The role of atmospheric nitrogen seems not to be well understood in NPS studies in China. This could have a major impact on policy decisions regarding the presumed agricultural source of nitrogen in ambient waters. At the very least, unit area estimates need to be developed for atmospheric nitrogen for all of agricultural and urban China so that this can be deducted from in-stream calculations of NPS-nitrogen when calculating the agricultural and rural contribution to pollution.

Hydrological observation system

China needs an experimental hydrological observation system that focuses on small catchments and on single types of land use. In North America, this was a research network that eventually covered most types of land uses and allowed coordinated hydrological and water quality studies that could be explicitly linked to specific types of land uses. This would require significant support from organizations such as the Ministry of Water Resources and Ministry of Agriculture to ensure that the network is developed around common principles and methods.

Best management practices (BMPs)

Policy makers need to focus on defining best management practices that will effectively reduce important non-point sources and methods for transferring these to local agricultural communities. In North America, agriculture departments provide broad-based extension services to the farming community, whereas in China much of this is available to farmers only on a fee-for-service basis. Clearly, this has a negative impact on the environmental behaviour of impoverished small farmers in China and requires a change in policy by the Ministry of Agriculture.

An example of the need for BMPs is in the disposal of animal manure. In China, manure (excluding feed lots) management is variable and excreta is also used as nutrient in fish ponds. In past decades, urban human excreta was applied to fields but this has been discontinued for a variety of social and economic reasons. Nevertheless, the differences in manure management between China and North America need to be understood, and best practices developed for manure management in China within the context of economic benefits for farmers in terms of reduced fertilizer costs.

National inventory of NPS data

While there is a growing literature of empirically based information, there is no national inventory of NPS results. Except for published papers, there is no mechanism through which researchers can pool their results as a means of developing a national NPS database of peer-reviewed data. In other countries such as the US, national agencies such as the USEPA and the US Department of Agriculture facilitate integration of studies and promote national coordination of information. There appears to be no such mechanism in China, either from the MEP or the Ministry of Agriculture.

Governance

A number of the issues noted above reflect the particular governance situation in China in which government ministries at all levels have little horizontal integration or coordination and therefore often have difficulty in finding common ground between themselves. This vertical structure also extends to research institutes and university groups which tend to have far less communication between them than among similar organizations in most western countries. This reflects a historical cultural phenomenon in China of vertical 'power' structures which is difficult to change quickly. The State Council of China recognizes this problem and is taking measures to realign government functions to better integrate overlapping or competing interests. The March 2008 decision to elevate SEPA to ministerial status may lead to greater leadership by the MEP in areas such as NPS pollution; however, western experience suggests that it takes more than 20 years to move from a vertical 'command-and-control' structure to one of horizontal coordination and facilitation among ministries. In the meantime, it is likely to require agencies such as the National Development and Reform Commission that provides overall guidance to the Chinese government on economic and reform issues, to enhance horizontal cooperation through direct intervention in pollution mitigation as, for example, the current focus on the restoration of Lake Tai (Taihu) in which NPS pollution is believed to be of major significance.

In conclusion, there is broad recognition of the role of agriculture and other types of NPSs in water pollution in China. There is yet, however, no consensus on the importance of NPSs relative to point sources. The policy responses reflect the absence of definitive information and have focused mainly on the more egregious examples of NPSs such as animal feedlots. However, in China, the gap between policy/legislation and enforcement remains large and is unlikely to change quickly despite the national government's efforts to create a more accountable system of evaluating officials' performance in environmental control. Policy also needs to be followed up by an expanded effort in educating local officials about NPS pollution, by improved monitoring and reporting of point sources as a basis of determining the real point source contribution, and by a more proactive farm-oriented programme of training in best management practices. It is unlikely that China will have a comprehensive approach to farm-based pollution for a considerable length of time due to the pressure of meeting other priorities in implementing its 'socialist countryside' initiative that aims to improve the life and livelihood of the rural population.

Acknowledgements

Dr William Booty kindly reviewed the draft manuscript and provided insight on certain technical procedures of the SWAT Model. Professor Jose Albiac provided constructive comments that improved the draft manuscript. The draft manuscript was also reviewed by three anonymous Chinese referees who made helpful comments that have been incorporated into the present text. Insight into Chinese NPS studies gained by the senior author (Ongley) through his involvement in the MEP/MWR/World Bank GEF project on Hai Basin Integrated Water and Environment Management is gratefully acknowledged.

Notes

1   The basin was not named due to confidentiality of the study.
2   Unpublished draft report by Chinese Research Academy of Environmental Sciences.
3   A sixth method using N and P input-output accounting procedures at the farm level is not currently used in China to our knowledge.
4   Personal communication, Professor Xia (Ministry of Environmental Protection).

References

ADB (2004) *Study on Control and Management of Rural Non-point Source Pollution*, Final Report of TA No. 3891-PRC, Asian Development Bank, Manila

ADB (2005) *Songhua River Basin Water Quality and Pollution Control Management – Summary Report*, Technical Assistance Consultant's Report to ADB, Asian Development Bank, Manila

ADB (2008) 'ADB invests in China wastewater treatment industry', posted 15 April 2008, http://afp.google.com/article/ALeqM5g3OOPaGeWlH38p9YbF7rULvNIcRQ, accessed October 2008

AERC (undated) 'The AERC National Export Coefficient Model', Aquatic Environments Research Centre, University of Reading, Reading, www.defra.gov.uk/farm/environment/water/csf/pdf/landuse-ges-append.pdf, accessed 15 November 2007

Agrawal, G. D., Lunkad, S. K. and Malkhed, T. (1999) 'Diffuse agricultural nitrate pollution of groundwaters in India', *Water, Science and Technology*, vol 29, no 3, pp67–75

Benjamin, L. P., Nina, F. C. and Mihael, L. P. (1991) 'Human influence on river nitrogen', *Nature*, vol 350, no 4, pp386–387

Bouraoui, F. and Dillaha, T. A. (1996) 'ANSWERS-2000: Run-off and sediment transport model', *Journal of Environmental Engineering*, ASCE, vol 122, no 6, pp493–502

Bouraoui, F. and Dillaha, T. A. (2000) 'ANSWERS-2000: Non-point source nutrient transport model', *Journal of Environmental Engineering*, ASCE, vol 126, no 11, pp1045–1055

Cai, M., Li, H. E. and Zhuang, Y. T. (2005) 'Rainfall deduction method for estimating non-point source pollution load for watershed', *Journal of Northwest Sci-Tech Univ of Agriculture and Forestry*, vol 33, no 4, pp102–106 (in Chinese)

Chen, J., He, D. and Cui, S. (2003) 'The response of river water quality and quantity to the development of irrigated agriculture in the last four decades in the Yellow River Basin, China', *Water Resources Research*, vol 39, no 3, pp1047

Chen, Y. Y., Hui, E. Q. and Jin, C. J. (2003) 'A hydrological method for estimation of non-point source pollution loads and its application', *Research of Environmental Sciences*, vol 16, no 1, pp10–13 (in Chinese)

Cheng, B., Zhang, Z. and Chen, L. (2005) 'Eutrophication of Taihu Lake and pollution from agricultural non-point sources in Lake Taihu Basin', *Journal of Agro-Environment Science*, vol 24, pp118–124 (in Chinese)

Cheng, H. G., Hao, F. H. and Ren, X. Y. (2006a) 'The study of the rate loss of nitrogenous non-point source pollution loads in different precipitation levels', *Acta Scientiae Circumstantiae*, vol 26, no 3, pp392–397 (in Chinese)

Cheng, H. G., Yue, Y. and Yang, T. S. (2006b) 'An estimation and evaluation of non-point source (NPS) pollution in the Yellow River Basin', *Acta Scientiae Circumstantiae*, vol 26, no 3, pp384–391 (in Chinese)

Cui, G., Zaheer, I. and Luo, J. (2003) 'Quantitative evaluation of non-point pollution of Taihu watershed using geographic information system', *Journal of Lake Sciences*, vol 15, no 3, pp236–244

Ding, X.-J., Yau, Q. and Ruan, X.-H. (2003) 'Waste load model for the Taihu Basin', *Advances in Water Science*, vol 14, no 2, pp189–192

Donigian, A. S. and Crawford, N. H. (1976) *Modelling Non-point Pollution from the Land Surface*, Hydrocomp Inc., USEPA-600/3-76-083

Duan, L., Duan, Z.-Q. and Xia, S.-Q. (2006) 'Quantification of non-point pollution for uplands in Taihu Lake catchment', *Bulletin of Soil and Water Conservation*, vol 26, no 6, pp40–43 (in Chinese)

Endreny, T. A. and Wood, E. F. (2003) 'Watershed weighting of export coefficients to map critical phosphorous loading areas', *Journal of the American Water Resources Association*, vol 39, no 1, pp165–181

Foster, G. R., Lane, L. J., Nowlin, J. D., Laflen, J. M. and Young, R. A. (1980) 'A model to estimate sediment yield from field-size areas: Development of model', in W. G. Knisel (ed) *CREAMS: A Field-scale Model for Chemicals, Run-off, and Erosion from Agricultural Management Systems*, US Department of Agriculture, Science and Education Administration, Chapter 3, pp36–64

Garland, E. J., Szydlik, J. J., Larson, C. E. and DiToro, D. M. (1999) 'Advanced Eutrophication Modeling of The Upper Mississippi River', Draft Final Report, HydroQual Inc. www.metrocouncil.org/Environment/RiversLakes/documents/LakePepinEutrophicationModel.pdf, accessed 30 November 2007

Goolsby, D. A., Battaglin, W. A., Lawrence, G. B., Artz, R. S., Aulenbach, B. T., Hooper, R. P., Keeney, D. R. and Stensland, G. J. (1999) *Flux and Sources of Nutrients in the Mississippi–Atchafalaya River Basin: Topic 3 Report for the Integrated Assessment on Hypoxia in the Gulf of Mexico*, NOAA Coastal Ocean Program Decision Analysis Series No 17, NOAA Coastal

Ocean Program, Silver Spring, MD, 130pp, (As cited in Gulf of Mexico Alliance White Paper Reductions in Nutrient Loading to the Gulf of Mexico 26 May 2005)

Han, F. P., Zhang, X. C. and Wang, Y. Q. (2006) 'The estimation on loading of non-point source pollution (N, P) in different watersheds of Yellow River', *Acta Scientiae Circumstantiae*, vol 26, no 11, pp1893–1899 (in Chinese)

Hao, F. H., Cheng, H. G. and Yang, S. T. (2006) *Non-point Source Model-Principle and Application*, Chinese Environmental Science Press, Beijing, pp24–28 (in Chinese)

Jin C. J., Li, H., and Cai, Y. (2004). 'Discussion on Investigation Method of Surface source Pollution in Songhuajiang River Basin', *Water Resource & Hydropower of Northeast*, vol 22, no 6, pp54–56 (in Chinese)

Johnes, P. J. (1996) 'Evaluation and management of the impact of land use change on the nitrogen and phosphorus load delivered to surface waters: The export coefficient modelling approach', *Journal of Hydrology*, vol 183, no 3–4, pp323–349

Liang, B., Wang, X. Y. and Cao, L. P. (2004) 'Water environment non-point source pollution loading estimation methods in China', *Journal of Jilin Normal University (Natural Science Edition)*, vol 3, pp58–62 (in Chinese)

Liu, F. (2000) 'Environmental Effects of the Beef Industry', Agricultural and Natural Resource Economics Discussion Paper 4/00, School of Natural and Rural Systems Management, University of Queensland, Australia, www.nrsm.uq.edu.au/discussionpapers/2000/ANREDP400.pdf, accessed 16 January 2008

Ma, X. B., Fok, L., Koenig, A. and Xue, Y. (2005) 'Estimation of total nitrogen export in the East River (Dongjiang) Basin. I. Export Coefficient Method', in J. H. W. Lee and K. M. Lam (eds) *Environmental Hydraulics and Sustainable Water Management*, Taylor and Francis Group, London, pp711–717

McElroy, A. D., Chiu, S. Y., Nebgen, J. W., Aleti, A. and Bennett, F. W. (1976) *Loadings Functions for Assessment of Water Pollution from Non-point Sources*, USEPA-600/2-76-151, May 1976

NC-DENR (1998) 'Development and Evaluation of Export Coefficient Methods and the Evaluation of the Nutrient Loss Evaluation Worksheet (NLEW) for Lower Coastal Plain Watersheds', North Carolina Department of Environment and Natural Resources, http://h2o.enr.state.nc.us/nps/nlew98.pdf, accessed 22 December 2007

Neitsch, S. L., Arnold, J. G., Kiniri, J. R., Srinivasan, R., and Williams, J. R. (2002) *Soil and Water Assessment Tool User Manual Version 2000*, Grassland, Soil and Water Research Laboratory and Blackland Research Center, Texas

Neitsch, S. L., Arnold J. G. and Kiniry, J. R. (2005) *Soil And Water Assessment Tool Theoretical Documentation*, Grassland, Soil and Water Research Laboratory, Texas, pp359–364

Ongley, E. D. (1987) 'Scale effects in fluvial sediment-associated chemical data', *Hydrological Processes*, vol 1, pp171–179

Ongley, E. D. (1996) 'Control of Water Pollution from Agriculture', FAO Irrigation and Drainage Paper 55, Food and Agriculture Organization of the United Nations, Rome

Shuai, H. and Xia, B.-C. (2006) 'Effects of land use structure on non-point source pollution in the areas of Guangzhou-Foshan', *Tropical Geography*, vol 26, no 3, pp229–233 (in Chinese)

Smith, R. A., Schwarz, G. E. and Alexander, R. B. (1997) 'Regional interpretation of water-quality monitoring data', *Water Resources Research*, vol 33, no 12, pp2781–2798

USDA (1976) *Control of Water Pollution from Cropland*, Agricultural Research Service, US Department of Agriculture, Report No ARS-H-5-2 (co-published with USEPA), Washington, DC

USDA (1983) *National Engineering Handbook: Sedimentation*, Department of Agriculture, Soil Conservation Service, Washington, DC

USEPA (1979) *Non-Point Source–Stream Nutrient Level Relationships: A Nationwide Study*, USEPA-600/3-79-105, Corvallis, Oregon, US

USEPA (1996) 'BASINS (Non-point source model)', www.epa.gov/OST/BASINS, accessed 12 February 2007

USEPA (2001) 'PLOAD User's Manual, Version 3', www.epa.gov/waterscience/BASINS/b3docs/PLOAD_v3.pdf, accessed 12 February 2007

USEPA (2003a) *National Management Measures for the Control of Non-point Pollution from Agriculture*, US Environmental Protection Agency Office of Water, Washington, DC

USEPA (2003b) *National Pollutant Discharge Elimination System Permit Regulation and Effluent Limitation Guidelines and Standards for Concentrated Animal Feeding Operations (CAFOs); Final Rule*, Federal Register Washington, DC

Uttormark, P. D., Chapin, J. D. and Green, K. M. (1974) *Estimating Nutrient Loadings of Lakes from Non-point Sources*, USEPA-600/3-74-020, August 1974, Washington, DC

Wang, S. P., Yu, L. Z. and Xu, S. Y. (2002) 'Research of non-point source pollution loading in Suzhou Creek', *Research of Environmental Sciences*, vol 15, no 6, pp20–24 (in Chinese)

Wischmeier, W. H. (1976) 'Use and misuse of the universal soil loss equation', *Journal of Soil and Water Conservation*, vol 31, pp5–9

Young, R. A., Onstad, C. A., Bosch, D. D. and Anderson, W. P. (1994) 'Agricultural Non-point Source Pollution Model, Version 4.03', AGNPS User's Guide, USDA-NRS-NSL, Oxford, MS, July 1994

Zhang, W. L., Wu, S. X. and Ji, H. J. (2004) 'Estimation of agricultural non-point source pollution in China and the alleviating strategies', *Scientia Agricultura Sinica*, vol 37, no 7, pp1008–1017 (in Chinese)

Zheng, Y. and Wang, X. J. (2002) 'Advances and prospects for non-point source studies', *Advances in Water Science*, vol 13, no 1, pp105–110 (in Chinese)

3

Integration of Modelling, Research and Monitoring in the Chesapeake Bay Program

Lewis Linker, Gary Shenk, Ping Wang and Richard Batiuk

Introduction

The Chesapeake Bay's 166,000km^2 watershed is in the eastern US and includes parts of New York, Pennsylvania, West Virginia, Delaware, Maryland and Virginia (Figure 3.1). Throughout the Chesapeake Bay watershed there are more than 100,000 streams and rivers that eventually flow into the Bay (USEPA, 2003a). Run-off and groundwater from the watershed flows into an estuary with a surface area of 12,000km^2 resulting in a land to water ratio of 14 to 1. This ratio is a key factor in explaining the significant influence the watershed has on Chesapeake Bay water quality. The nine major basins of the Chesapeake Bay watershed are the Susquehanna, Potomac, Patuxent, Rappahannock, York and James Rivers and the Maryland Western Shore, Maryland Eastern Shore and Virginia Eastern Shore (Figure 3.1, Plate 1).

The Susquehanna is the largest river, followed by the Potomac and James Rivers. The current land use in the watershed (Figure 3.2, Plate 2) is about 65 per cent forest or wooded, 24 per cent agriculture and 11 per cent developed land (buildings, roads and so on, in urban, suburban and rural areas). Nearly 16 million people live in the Chesapeake Bay watershed.

To simulate the Chesapeake watershed, the river flows and associated transport and fate of nutrients and sediment, and the effects of these loads on the water quality of the Chesapeake, the integrated models of the Chesapeake Bay were developed which are a set of interactive models of the airshed, watershed,

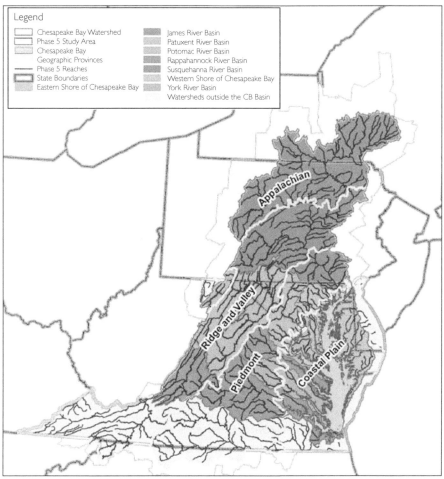

Figure 3.1 *Phase 5 study area showing major watersheds, rivers and geographic provinces*

estuary, living resources and climate change directed towards regional Bay water quality issues.

The Chesapeake Bay Program (CBP) has been applying increasingly sophisticated integrated models for more than two decades. The first integrated models were relatively crude, being nothing more than a simple linkage of a watershed model and a model of the estuary. As the scope and sophistication of decision making grew in the Chesapeake, commensurate with increased challenges posed by the increased population and growth in the region, the integrated models have developed to include models of the airshed, watershed, estuary, living resources and climate change.

The CBP integrated models were tailored to the nature and scale of the basic questions of the CBP decision makers (Chesapeake Executive Council, 1987; Koroncai et al, 2003) which were:

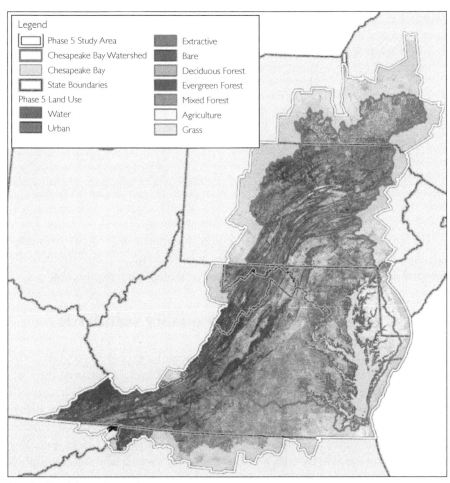

Note: See Plate 2 for a colour version.

Figure 3.2 *Phase 5 study area showing states in the watershed and major land uses*

1   What input levels of nitrogen, phosphorus and sediment from the watershed and airshed will achieve the Chesapeake water quality standards?
2   What is the magnitude of the different point and non-point source nutrient loads and how do they compare among all sources including atmospheric pollutant sources?
3   What are cost-effective and equitable nutrient and sediment reductions among the jurisdictions considering the different loads from different media and states?

CBP decision makers also want the management of local streams and watersheds to be efficiently integrated into the larger regional Bay water quality standards. For this reason, the integrated CBP models are scaled to support local small-scale model needs though a community model approach of web-shared

Table 3.1 *Annual average Chesapeake basin-wide nutrient and sediment tributary strategy caps compared with estimated all-forest loads (lowest nutrient loads to the Bay), estimated 1985 loads (highest loads to the Bay) and estimated year 2000 and 2007 conditions*

	Estimated all-forest loads	Estimated 1985 loads	Estimated 2000 loads	Estimated 2007 loads	Estimated tributary strategy cap loads
Total nitrogen	27.5	153.1	129.2	118.8	79.4
Total phosphorus	0.5	12.3	8.7	8.3	5.8
Total suspended sediment	1.40	5.29	4.58	4.32	3.76

Note: Units of total nitrogen and phosphorus in million kilograms. Units of total suspended sediment in million metric tons.

model code and input data (http://ccmp.chesapeake.org/CCMP/models/CBPhase5/index.php). This allows more cost-effective and environmentally protective decisions to be made at local, state, regional and federal scales.

Chesapeake Bay water quality standards

To achieve and maintain the water quality conditions necessary to protect the aquatic living resources of the Chesapeake Bay and its tidal tributaries, the CBP developed water quality criteria for dissolved oxygen, clarity and chlorophyll (USEPA, 2003b). These published criteria, along with criteria attainment assessment procedures and refined tidal water designated uses (USEPA, 2003a), were adopted by the states into their water quality standards. The Chesapeake Bay water quality standards, based on dissolved oxygen, water clarity and chlorophyll a, are an integrated set of criteria that provide the basis for defining the water quality conditions necessary to protect Chesapeake Bay aquatic living resources from effects of nutrient and sediment over-enrichment (Koroncai et al, 2003; USEPA, 2003b). Reductions in the watershed of nutrient loads by about a half and sediment loads by about a third from the 1980s zenith of pollutant loads (Table 3.1) are necessary to achieve and maintain these water quality criteria (Koroncai et al, 2003).

Nutrient cap load allocations

The effect of nutrient loads on water quality and living resources tends to vary considerably by season and region. Low dissolved oxygen problems tend to be more pronounced in the deeper parts of the upper-Bay region during the summer months. The allocations for nutrients were developed primarily to address this problem. The allocations for sediment were primarily directed towards the restoration and protection of Chesapeake underwater grasses.

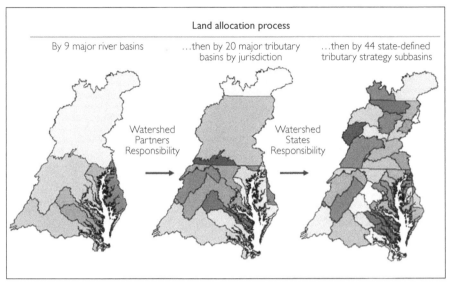

Note: See Plate 3 for a colour version.

Figure 3.3 *The process used to allocate nutrient and sediment load reductions and caps to first the major basins, then the state-basins, then sub-basins within the state-basin*

As a result, the states and the US Environmental Protection Agency (USEPA) agreed to cap average annual nitrogen loads delivered to the Bay's tidal waters at 79.4 million kilograms and average annual phosphorus loads at 5.8 million kilograms based on the findings of the integrated CBP models corroborated by CBP monitoring and research (Koroncai et al, 2003). It is estimated that these allocations will require reductions, from 2000 levels, in nitrogen pollution by 50 million kilograms and phosphorus pollution by 2.9 million kilograms. Sediment loads delivered to the Bay's tidal waters are capped at 3.76 million metric tons/year (Table 3.1).

The CBP partners, consisting of the states and the federal government, agreed to these load reductions on the basis of the integrated airshed, watershed and Bay water quality models (Cerco and Noel, 2004). The integrated models projected the nutrient and sediment load reductions (Table 3.1) required to attain the dissolved oxygen criteria and significantly reduce the persistent summer anoxic conditions in the deep, bottom waters of Chesapeake Bay restoring suitable habitat quality conditions throughout the tidal tributaries (Koroncai et al, 2003). Furthermore, these reductions are projected to eliminate excessive, sometimes harmful, algae conditions (measured as chlorophyll a) throughout the Chesapeake. The six CBP states agreed to the allocation of the basin-wide cap loads for nitrogen, phosphorus and sediment by major tributary basin and jurisdiction (Koroncai et al, 2003). The basin-wide cap loads became the basis for tributary strategies developed by each of the states to reduce their nutrient and sediment loads to meet their caps (Figure 3.3, Plate 3, Table 3.2).

Table 3.2 *Pollution reduction amounts and costs by state (jurisdiction)*

Jurisdiction	Pollutant	2002 load	Cap allocation	Needed reduction from 2002 level	Estimated tributary strategy costs
Virginia	Nitrogen	35.29	23.31	11.97	
	Phosphorous	4.46	2.72	1.74	
	Sediment	2.16	1.76	0.40	
					$6.3 billion
Maryland	Nitrogen	25.72	16.90	8.82	
	Phosphorous	1.80	1.32	0.48	
	Sediment	0.92	0.65	0.27	
					$9.0 billion
Pennsylvania	Nitrogen	49.53	32.61	16.92	
	Phosphorous	1.62	1.03	0.60	
	Sediment	1.00	0.90	0.10	
					$8.6 billion
New York	Nitrogen	8.26	5.72	2.54	
	Phosphorous	0.46	0.27	0.19	
	Sediment	0.13	0.12	0.01	
					$0.45 billion
West Virgiinia	Nitrogen	3.24	2.15	1.08	
	Phosphorous	0.26	0.17	0.09	
	Sediment	0.31	0.29	0.02	
					$0.35 billion
Delaware	Nitrogen	2.28	1.31	0.97	
	Phosphorous	0.18	0.14	0.05	
	Sediment	0.05	0.04	0.01	
					$0.30 billion
District of Columbia	Nitrogen	1.62	1.09	0.53	
	Phosphorous	55,645.35	0.15	0.00	
	Sediment	0.01	0.01	0.00	
					$4.3 billion
Total					**$29.3 billion**

Note: Nitrogen and phosphorus loads in million kilograms, sediment load in million metric tons, and cost in US dollars.
Source: National Academy of Public Administration (2007)

The Watershed Model

The Watershed Model is based on the Hydrologic Simulation Program – Fortran (HSPF), a widely used public domain and open source code, and is very similar to the BASINS Model (Bicknell et al, 2001). The Watershed Model is developed as a community model addressing regional water quality in the Chesapeake, while still capable of addressing small-scale state water quality needs at a fine watershed scale.

There have been five previous versions of the Watershed Model over the past two decades and over time the Watershed Model refinements have tended towards increased segmentation, longer simulation periods, and greater land use

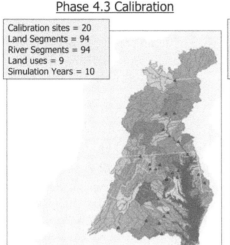
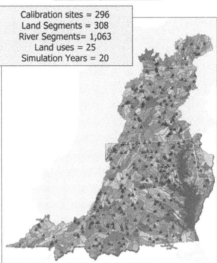

Phase 4.3 Calibration

Calibration sites = 20
Land Segments = 94
River Segments = 94
Land uses = 9
Simulation Years = 10

Phase 5 Calibration

Calibration sites = 296
Land Segments = 308
River Segments= 1,063
Land uses = 25
Simulation Years = 20

Note: See Plate 4 for a colour version.

Figure 3.4 *Comparison of previous Watershed Model phase and Phase 5
segmentation and calibration stations*

and best management practices (BMPs) mechanistic detail (Donigian et al, 1994; Linker, 1996; Linker et al, 1996, 2000; USEPA, 2009). The most recent version, Phase 5, increases the segmentation to about 1000 model segments at an average size of about 170km^2 (Figure 3.4, Plate 4). This allows greater application of calibration stations, of which 296 are used for the calibration of hydrology or other water quality components – an increase of an order of magnitude compared with the Chesapeake watershed stations used for the previous Watershed Model version. Increased river-reach segmentation resulted in a 12-fold increase in monitoring stations, and improved characterization of spatial variation of the river reaches (within the limitations of the completely mixed reaches of the HSPF code). In Phase 5, the model simulation period runs 21 years from 1985 to 2005 to take advantage of recent and expanded monitoring. Land use is a time series input that changes annually over the simulation period.

Phase 5 has greater mechanistic detail including an expansion of land uses to 13 types of cropland, 2 types of woodland, 3 types of pasture, 4 types of urban land, and other special land uses such as surface mines and construction land uses detail (USEPA, 2009). Key model inputs of manures, fertilizers and atmospheric deposition of nutrients are on an annual time series, using a mass balance of Agricultural Census animal populations, crops, records of fertilizer sales, and other data sources, as well as daily point source loads. Non-point source BMPs change annually, have refined nutrient and sediment reduction efficiencies, and vary their efficiency on the basis of storm size.

Community model approach

The Chesapeake community model consists of open source, public domain programmes of model code, preprocessors, postprocessors and input data that are freely distributed often over the web: http://ccmp.chesapeake.org/CCMP/models/CBPhase5/index.php. The operating system, Linux, is also open source. Model input data, such as the precipitation fields, point source discharges, atmospheric deposition and land use are made freely available in a web-based data-sharing approach. The current Watershed Model, Phase 5, is specifically designed as a community model that can be used in a direct, *as-is* application, or can be used as a point of departure for more detailed, small-scale models. The data sharing and the modularity of Phase 5 are intended to encourage the efficient use of the model's data, or particular model elements, in other independent analyses or models of the watershed.

The use of the community model approach is adopted and expanded by some state environmental agencies that plan to use the Phase 5 model in a *nested* model approach. This will allow better coordination between the small-scale models in the watershed and the river basin-scale nutrients and sediment reductions needed to achieve the Bay's water quality standards. Overall, the nested approach should be more effective, cost efficient and equitable.

The Watershed Model is linked to two other models that together form a simulation system sufficient for attainment analysis of the Chesapeake Bay water quality standards of dissolved oxygen, clarity and chlorophyll (Figure 3.5, Plate

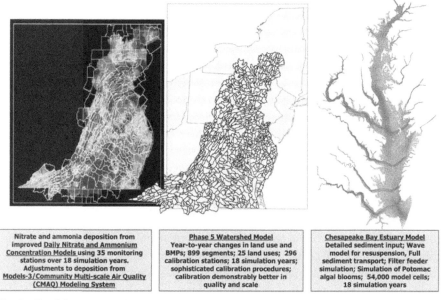

| Nitrate and ammonia deposition from improved Daily Nitrate and Ammonium Concentration Models using 35 monitoring stations over 18 simulation years. Adjustments to deposition from Models-3/Community Multi-scale Air Quality (CMAQ) Modeling System | Phase 5 Watershed Model Year-to-year changes in land use and BMPs; 899 segments; 25 land uses; 296 calibration stations; 18 simulation years; sophisticated calibration procedures; calibration demonstrably better in quality and scale | Chesapeake Bay Estuary Model Detailed sediment input; Wave model for resuspension, Full sediment transport; Filter feeder simulation; Simulation of Potomac algal blooms; 54,000 model cells; 18 simulation years |

Note: See Plate 5 for a colour version.

Figure 3.5 *Overview of the Chesapeake integrated models of the airshed, watershed and estuary*

5). They are the Airshed Model and the Water Quality and Sediment Transport Model (WQSTM) (also referred to as the Bay Model).

The Airshed Model

The Airshed Model, like the Watershed Model, is a loading model. The Airshed Model provides atmospheric deposition loads of nitrogen to the watershed land and water bodies including the tidal Bay and adjacent coastal ocean.

The Airshed Model is a combination of two models – a regression model of atmospheric wet deposition and a fully developed air simulation of the North American continent called the Community Multiscale Air Quality (CMAQ) Model (USEPA, 1999). The Airshed Model, like the other CBP models, has gone through a series of refinements with increasingly sophisticated regression and air quality models applied over time (Linker et al, 2000).

The regression model uses 34 National Atmospheric Deposition Program (NADP) monitoring stations and AirMoN stations (Figure 3.6, Plate 6) to form a regression of wetfall deposition in the entire Watershed Model domain over the entire simulation period (Grimm and Lynch, 2004). For each day of rain, a regression – using rainfall, land use and local emission levels of ammonia and nitrous oxides – estimates wetfall atmospheric deposition.

Dryfall deposition is continuous and the second Airshed Model – the CMAQ Model – estimates it daily. The CMAQ Model is run on a 36km grid covering the North American continent simulating boundary conditions with a refined 12km grid for the entire Phase 5 study area (Figure 3.7, Plate 7). In scenario mode, CMAQ also provides estimates of nitrogen deposition resulting from changes in emissions from utility, mobile and industrial sources due to management actions or growth. The base deposition that a regression determines is adjusted by a reduction ratio deposition determined by CMAQ.

The Water Quality and Sediment Transport Model

Taking the nutrient loads from the Airshed Model and the nutrient and sediment loads from the Watershed Model, the Bay Model, also called the WQSTM, is the decision model used to simulate water quality and living resource responses to the nutrient and sediment input loads. Together, these three models, along with ancillary models of key living resources, form the Chesapeake Bay integrated models.

Like the Watershed Model, the WQSTM has had several versions originating from simpler simulation systems. The first estuary model of the Chesapeake, completed in 1987, was a steady-state, three-dimensional simulation of the summer average period of 1965, 1984 and 1985 (USEPA, 1987). Increasingly sophisticated models followed with expanded spatial detail and simulation

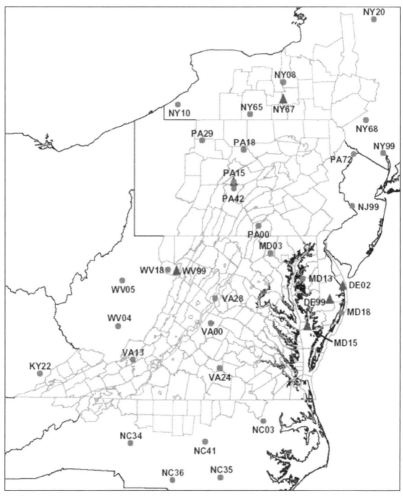

Note: See Plate 6 for a colour version.

Figure 3.6 *Atmospheric deposition monitoring stations used in the airshed regression model*

periods, and extension of the simulation to key living resources such as submerged aquatic vegetation (SAV) and oysters (Cerco and Cole, 1994; Thomann et al, 1994; Cerco and Meyers, 2000; Cerco and Noel, 2004).

The WQSTM is a three-dimensional model of the tidal Bay comprised 57,000 cells (Figure 3.8, Plate 8). which incorporates a full sediment transport simulation. The central issues of the WQSTM are computations of algal biomass, dissolved oxygen and water clarity. To compute algae and dissolved oxygen, a suite of 24 model state variables is necessary (Table 3.3).

The Chesapeake Bay Management Program (CBMP) includes other linked or coupled models. A hydrodynamic model simulates the hourly temperatures and movement of water in the Bay. The water and habitat quality response to

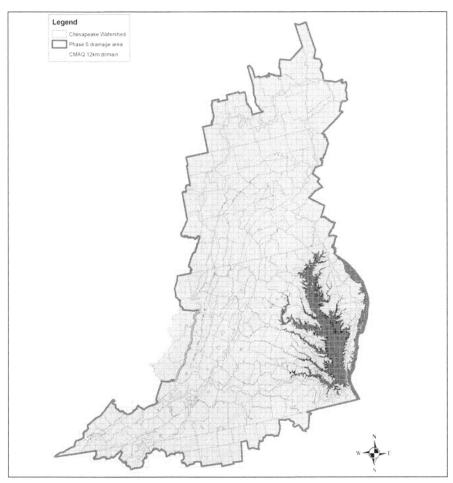

Note: See Plate 7 for a colour version.

Figure 3.7 *The 12km CMAQ model grid over the Chesapeake Bay basin used for Phase 5 Model applications*

nutrient and sediment loads is simulated through coupled models of sediment diagenesis, benthos and submerged aquatic vegetation. Loads are inputs from the Watershed Model, from direct atmospheric deposition to the surface of the Bay, and estimated loads from the ocean boundary. The Bay Model is applied in one continuous simulation period (1985–2005) to model transport, eutrophication processes and sediment–water interactions under various management scenarios designed to analyse the water quality and living resource responses to load reductions at all points in the Bay.

The details of the development of the hydrodynamic and water quality models and their calibration and sensitivity are presented in Cerco and Cole (1994), Cerco (2000), Cerco and Meyers (2000), Wang and Johnson (2000), Cerco and Moore (2001) and Cerco and Noel (2004).

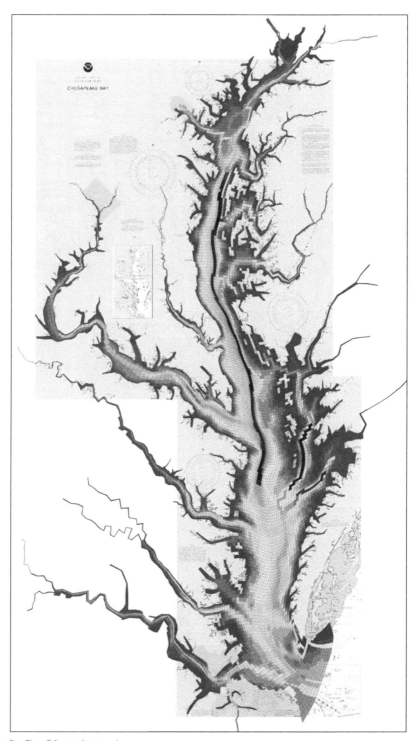

Note: See Plate 8 for a colour version.

Figure 3.8 *The 57,000 three-dimensional cells of the Bay Model*

Table 3.3 *WQSTM state variables*

Temperature	Dissolved organic nitrogen
Salinity	Labile particulate organic nitrogen
Inorganic suspended solids	Refractory particulate organic nitrogen
Diatoms	Total phosphate
Cyanobacteria (blue-green algae)	Dissolved organic phosphorus
Other phytoplankton	Labile particulate organic phosphorus
Dissolved organic carbon	Refractory particulate organic phosphorus
Labile particulate organic carbon	Dissolved oxygen
Refractory particulate organic carbon	Chemical oxygen demand
Ammonium	Dissolved silica
Nitrate	Particulate biogenic silica
Microzooplankton	Mesozooplankton

Source: Cerco and Noel (2004)

The WQSTM treats each cell as a control volume, which exchanges material with its adjacent cells. The WQSTM solves, for each volume and for each state variable, a three-dimensional conservation of mass equation (Cerco and Cole, 1994). The numerous details of the kinetics portion of the mass-conservation equation for each state variable are described in Cerco and Cole (1994) and Cerco and Noel (2004). The processes and phenomena relevant to the water quality model simulation include:

- bottom-water hypoxia;
- the spring phytoplankton bloom;
- nutrient limitations;
- sediment–water interactions; and
- nitrogen and phosphorus budgets.

Over seasonal time scales, sediments are a significant source of dissolved nutrients to the overlying water column. The role of sediments in the system-wide nutrient budget is especially important in summer when seasonal low flows diminish riverine nutrient input, sediment oxygen increases with warmer temperatures, and low dissolved oxygen causes large fluxes of ammonia and phosphate from the sediment. The WQSTM is coupled directly to a predictive benthic–sediment model (DiToro and Fitzpatrick, 1993). These two models interact at each time step with the WQSTM delivering settled organic material to the sediment bed and the benthic–sediment model calculating the flux of oxygen and nutrients to the water column.

The ultimate aim of eutrophication modelling is to preserve living resources. Underwater grasses, or SAV, are an important living resource because they provide a habitat for biota of economic importance and help support the estuarine food chain (USEPA, 2000). Establishing healthy SAV acres is also directly tied to the clarity water quality standard. The WQSTM's direct simulation of SAV accounts for the relationships among grass production, light and nutrient availability, allowing an estimate to be made of the response of SAV to reductions in nutrient and sediment loads. A thin ribbon of model cells following the 2m depth contour along the shore is the key littoral zone available for SAV growth. The SAV component of the model builds on the concepts established by Wetzel and Neckles (1986) and Madden and Kemp (1996).

Three state variables are modelled for SAV: shoots, (above-ground biomass), roots (below-ground biomass) and epiphytes (attached growth to leaves). In addition, the estuary model incorporates three dominant SAV communities based largely on salinity regimes (Moore et al, 1999). Within each community, a target species is selected: eelgrass (*Zostera marina*) for high salinity, widgeon grass (*Ruppia maritima*) for moderate salinity and wild celery (*Vallisneria americana*) for tidal fresh. Because SAV production in the Bay and tributaries is largely determined by light availability (Kemp et al, 1983; Orth and Moore, 1984), a predictive representation of light attenuation is needed. The computation of light attenuation requires the addition of fixed solids, or suspended sediment, to the list of model state variables.

In addition to simulating SAV as a living resource, the model simulates three phytoplankton groups (diatoms, greens and blue-greens) and separates zooplankton into two size classes for modelling purposes: microzooplankton (44–201μm) and mesozooplankton (>201μm). Zooplankton are selected as a parameter because they are a valuable food source for finfish and to improve the computation of phytoplankton since zooplankton feed on phytoplankton, detritus and each other.

Benthos, or bottom-dwelling organisms, are included in the model because they are an important food source for crabs, finfish, and other economically significant biota and because they can exert a substantial influence on water quality through their filtering of overlying water (Cohen, 1984; Newell, 1988). Within the estuary model, the benthos are divided into deposit feeders and filter feeders.

Integration of modelling, monitoring and research

Underlying all of the Chesapeake integrated models are monitoring data. Indeed, when monitoring and modelling programmes are done well, the integration of monitoring data and models is complete. No credible model can be developed without calibration to observed data. By the same token, monitoring programmes without supporting models are also incomplete, as modelling data allow the filling in and explanation of the discrete observations, both over time and between monitoring stations. Research plays a key role here as well. At different times, the integrated models instigated new research programmes in order to better understand nutrient fluxes in sediment, interactions and linkages between living resources and water quality, and sediment transport mechanisms. Often the research led to development of new monitoring or modelling approaches. The triad of modelling, monitoring and research are the three legs of the support underlying all successful integrated modelling programmes.

In the CBP, it was found that the integrated modelling led to integrated decision making as the models facilitated dialogue among the decision makers as they explored options in the range of meaningful, protective and equitable reductions and the degree of water quality and living resource improvements brought about by different levels of pollutant reduction. The emphasis was on understanding, not on the numbers generated by the models. Questions of how to compare the cost and effectiveness of nitrogen reduction from air emissions, point sources and non-point sources were examined. In the examination of different options, decision makers also came to understand the essential concerns and perspectives of other decision makers from the different jurisdictions of the six-state region of the Chesapeake watershed.

Case study: An application of climate change assessment in the Chesapeake

To estimate climate change in the Chesapeake watershed and estuary we examined the flows and associated nutrient and sediment loads in several major river basins of the Chesapeake Bay using the integrated Chesapeake models. Nine climate change scenarios were evaluated reflecting the range of potential changes in temperature and precipitation in the year 2030 based on projections from seven global climate models, two Intergovernmental Panel on Climate Change emission scenarios, and three assumptions about precipitation intensity in the largest events (CARA, 2007). Weather data reflecting each climate change scenario were created by modifying a 16-year period of historical data of precipitation and temperature from 1984 to 2000 (Linker et al, 2007). Climate change estimates were combined with a 2030 estimated land use based on a sophisticated land use model containing socio-economic estimates of development throughout the watershed. The Phase 5 Watershed Model was used for this climate change assessment. The assessment was supported by use of tools developed for USEPA's BASINS 4 system including the Climate Assessment Tool (CAT). Key basins of the Chesapeake watershed were examined and differences among the basin responses to future climate change were noted through a comparison with a base scenario without the estimated effects of climate change.

In our Chesapeake watershed, the 2030 year estimates of mean annual air temperature are for increases of about 1.5°C with a high degree of certainty. Estimated precipitation increase among the seven global climate models used are about 2 per cent, especially at higher rainfall events, and this is estimated with a low-to-moderate degree of certainty. How temperature and precipitation changes affect flow and associated nutrient and sediment loads in the watershed hangs in a hydrologic balance between precipitation and evapotranspiration. Temperature increases tend to increase evapotranspiration in watersheds and this can offset increases in precipitation.

This seems to be the case in the Chesapeake watershed. Current estimates of the medians of the nine different scenarios run have an annual average flow, nitrogen and phosphorus load decreases of –6.0 per cent, –1.6 per cent, and –2.1 per cent, respectively. Because sediment loads increase with higher rainfall events, the median of the nine scenario estimates for sediment is for an increase of 4.9 per cent.

This work will continue to be refined over the next several years as improved global climate models and downscaling techniques become available. The climate is changing, and this has significant implications for our long-term CBP goals.

Conclusion

The CBP integrated models will be applied over the three years from 2008 to 2010 to examine the measures needed to achieve water quality standards in 2010 and in future Chesapeake land use and populations of 2020 or 2030 (Figure 3.9, Plate 9). The decisions made will influence plans at the federal, state and local level. Providing tools to aid decision making at each of these scales is the objective of the Chesapeake integrated models.

The continued progress in nutrient and sediment reductions as described in Table 3.1 can be cited as evidence that the CBP's voluntary approach is working. The success to date in reducing nutrient and sediment loads has been gained despite continued robust human and animal population growth, as well as strong economic growth in the watershed in recent decades. Unfortunately, the pace of reductions is too slow to reach our goal of achieving fully implemented tributary strategy reductions by 2010.

This inadequate pace of Chesapeake restoration has caused the CBP to turn towards a more regulatory approach that will apply additional controls on point sources, concentrated animal feeding operations (CAFOs), and municipal separate stormwater sewer systems (MS4s) by the year 2011. While non-point sources will continue to be unregulated, reasonable assurances that non-point source load reduction measures will actually be applied will also be required.

Whether these additional regulatory levers are sufficient to increase the pace in and of themselves remains to be demonstrated, but the experience in the Bay Program so far is that the regulated discharges are among the loads with the highest rates of nutrient reduction. This is true of point source dischargers to both the air and water. Air point source dischargers are usually power plants, also called electric generating units (EGUs). The EGUs are covered under a separate regulatory authority, the Clean Air Act. Compared with 1980, atmospheric emissions from EGUs will be reduced by about 70 per cent for nitrogen, which combined with other Clean Air Act regulatory controls approximates a reduction of nitrogen loads delivered to the Bay by 2010 of about 8 million kilograms.

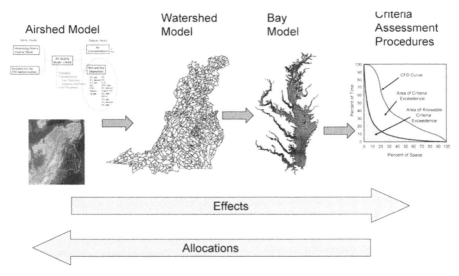

Note: See Plate 9 for a colour version.

Figure 3.9 *The CBP integrated models of the airshed, watershed, estuary water quality and sediment transport, key living resources, and climate change*

Similarly, point source reductions to date since 1985 have been reduced by 39 per cent for nitrogen and 62 per cent for phosphorus, reductions of 15.6 million and 2.6 million kilograms, respectively, of loads delivered to the Bay. This is compared with voluntary non-point source control measures that have achieved (in agricultural, urban, mixed open and septic loads) roughly a 20 per cent reduction in nitrogen and a 16 per cent reduction in phosphorus over the same period, equivalent to an 18 million kilogram nitrogen reduction and a 1 million kilogram phosphorus reduction in loads delivered to the Bay.

As we have developed this evolving mix of regulatory and voluntary nutrient and sediment controls in the CBP, we have learned that integrated modelling and decision making are necessary for progress, given the high cost of controlling multiple nutrient and sediment load sources in the Chesapeake. Further development and application of integrated models in the Chesapeake region is anticipated because the environmental control costs are high, and the simulation, tracking and management of the different pollutant sources in the different media of air and water, as well as among the different jurisdictions, allows decision making that is most cost effective, and environmentally protective. Integrated modelling and decision making provides for opportunities for greater efficiencies across environmental programmes in air, water and living resource management.

The control measures in the Chesapeake are complex and involve not only different media (air, water, living resources) but many distributed sources as well. We know that success in the Chesapeake can only be assured though engagement of all available pollutant control measures. The Chesapeake region has nutrient controls in the airshed and watershed that affect, in one way or another, everyone, doing everything, everywhere in the watershed.

Most importantly, application of the integrated models fosters dialogue among the six states and the federal government on how to appropriately share responsibility for nutrient and sediment reductions, and provides a level of understanding of the overall ecosystem of the airshed, watershed and Bay. The integrated models also support the tracking of nutrient and sediment controls allowing all to plan for reductions to offset future growth in the region and to encourage forward-looking momentum in the implementation of environmental protection.

References

Bicknell, B. R., Imhoff, J. C., Kittle, J. L. Jr, Jobes, T. H. and Donigian, A. S. Jr (2001) *Hydrologic Simulation Program – Fortran*, HSPF Version 12 Users Manual, US Environmental Protection Agency, Office of Research and Development, Athens, GA, and US Geological Survey, Hydrologic Analysis Software Support Program, Reston, VA

CARA (2007) Consortium for Atlantic Regional Assessment, www.cara.psu.edu/climate/7models_summary.asp, accessed 1 February 2008

Cerco, C. (2000) 'Phytoplankton kinetics in the Chesapeake Bay Eutrophication Model', *Water Quality and Ecosystem Modeling*, vol 1, no 1–4, pp5–49

Cerco, C. and Cole, T. (1994) *Three-Dimensional Eutrophication Model of Chesapeake Bay*, Technical Report EL-94-4, US Army Corps of Engineers Waterways Experiment Station, Vicksburg, MS

Cerco, C. and Meyers, M. (2000) 'Tributary refinements to Chesapeake Bay Model', *Journal of Environmental Engineering*, vol 126, no 2, pp164–174

Cerco, C. and Moore, K. (2001) 'System-wide submerged aquatic vegetation model for Chesapeake Bay', *Estuaries*, vol 24, no 4, pp522–534

Cerco, C. F. and Noel, M. R. (2004) *The 2002 Chesapeake Bay Eutrophication Model*, EPA 903-R-04-004, US Environmental Protection Agency, Chesapeake Bay Program Office, Annapolis, MD

Chesapeake Executive Council (1987) *Chesapeake Bay Agreement*, Chesapeake Executive Council, Annapolis, MD

Cohen, A. S. (1984) 'Effect of zoobenthic standing crop on laminae preservation in tropical lake sediment, Lake Turkana, East Africa', *Journal of Paleontology*, vol 58, pp499–510

DiToro, D. M. and Fitzpatrick, J. J. (1993) *Chesapeake Bay sediment flux model*, Contract Report EL-93-2, US Army Engineer Waterways Experiment Station, Vicksburg, MS, NTIS No AD A267 189

Donigian, A. S. Jr, Bicknell, B. R., Patwardhan, A. S., Linker, L. C., Chang, C. H. and Reynolds, R. (1994) *Chesapeake Bay Program Watershed Model Application to Calculate Bay Nutrient Loadings*, US Environmental Protection Agency, Chesapeake Bay Program Office, Annapolis, MD

Grimm, J. W. and Lynch, J. A. (2004) 'Enhanced wet deposition estimates using modeled precipitation inputs', *Environmental Monitoring and Assessment*, vol 90, no 1–3, pp243–268

Kemp, W. M., Twilly, R. R., Stevenson, J. C., Boynton, W. R., and Means, J. C. (1983) 'The decline of submerged vascular plants in Upper Chesapeake Bay: Summary of results concerning possible causes', *Marine Technology Society Journal*, vol 17, no 2, pp78–89

Koroncai, R., Linker, L., Sweeney, J. and Batiuk, R. (2003) *Setting and Allocating the Chesapeake Bay Basin Nutrient and Sediment Loads*, US Environmental Protection Agency, Chesapeake Bay Program Office, Annapolis, MD, www.chesapeakebay.net/pubs/doc-allocating-whole.pdf, accessed 3 August 2007

Linker, L. C. (1996) 'Models of the Chesapeake Bay', *Sea Technology*, vol 37, no 9, pp49–55

Linker, L. C., Stigall, C. G., Chang, C. H. and Donigian, A. S. Jr (1996) 'Aquatic accounting: Chesapeake Bay Watershed Model quantifies nutrient loads', *Water Environment and Technology*, vol 8, no 1, pp48–52

Linker, L., Shenk, G., Dennis, R. and Sweeney, J. (2000) 'Cross-media models of the Chesapeake Bay watershed and airshed', *Water Quality and Ecosystem Modeling*, vol 1, no 1–4, pp91–122

Linker, L. C., Johnson, T., Shenk, G. W., Cerco, C. F. and Wang, P. (2007) 'Evaluating 2030 climate change in the Chesapeake Bay watershed and estuary', *19th Biennial Conference of the Estuarine Research Federation*, 4–8 November 2007, Providence, RI

Madden, C. J. and Kemp, W. M. (1996) 'Ecosystem model of an estuarine submersed plant community: Calibration and simulation of eutrophication responses', *Estuaries*, vol 19, no 2B, pp457–474

Moore, K. A., Wilcox, D. J., Orth, R. J. and Bailey, E. (1999) *Analysis of Historical Distribution of Submerged Aquatic Vegetation (SAV) in the James River*, Report to Virginia Coastal Resources Management Program, Department of Conservation and Recreation, Richmond, VA.

National Academy of Public Administration (2007) *Taking Environmental Protection to the Next Level: An Assessment of the US Environmental Services Delivery System*, Report for the US Environmental Protection Agency, Washington, DC, www.napawash.org/pc_management_studies/EPA_FULL_Report_April2007.pdf, accessed 1 February 2008

Newell, R. I. E. (1988) 'Ecological changes in Chesapeake Bay: Are they the result of overharvesting the eastern oyster (*Crassostrea virginica*)? Understanding the estuary: Advances in Chesapeake Bay research', in E. Krome and M. Haire (eds) *Proceedings of March 1988 Chesapeake Bay Conference*, CRC Publication 129, Chesapeake Research Consortium, Edgewater, MD, pp536–546

Orth, R. J. and Moore, K. A. (1984) 'Distribution and abundance of submerged aquatic vegetation in Chesapeake Bay: An historical perspective', *Estuaries*, vol 7, pp531–540

Thomann, R. V., Collier, J. R., Butt, A., Casman, E. and Linker, L. C. (1994) *Response of the Chesapeake Bay Water Quality Model to Loading Scenarios*, CBP/TRS 101/94, US Environmental Protection Agency, Chesapeake Bay Program Office Annapolis, MD

USEPA (1987) *A Steady State Coupled Hydrodynamic/Water Quality Model of the Eutrophication and Anoxic Process in the Chesapeake Bay*, US Environmental Protection Agency, Chesapeake Bay Program Office Annapolis, MD, Work Assignment Number 40 EPA Contract Number 68-03-3319

USEPA (1999) *Science Algorithms of the EPA Models-3 Community Multiscale Air Quality (CMAQ) Modeling System*, EPA/600/R-99/030, www.epa.gov/asmdnerl/CMAQ/CMAQscienceDoc.html, accessed 3 August 2007

USEPA (2000) *Submerged Aquatic Vegetation Water Quality and Habitat-Based Requirements and Restoration Targets*, CBP/TRS 245/00 EPA 903-R-00-014, www.chesapeakebay.net/pubs/sav/index.html, accessed 3 August 2007

USEPA (2003a) *Technical Support Document for Identification of Chesapeake Bay Designated Uses and Attainability*, US Environmental Protection Agency, Chesapeake Bay Program Office Annapolis, MD, www.chesapeakebay.net/pubs/waterqualitycriteria/uaa08052003/uaa.pdf, accessed 3 August 2007

USEPA (2003b) *Ambient Water Quality Criteria for Dissolved Oxygen, Water Clarity and Chlorophyll a for the Chesapeake Bay and Its Tidal Tributaries* US Environmental Protection Agency,

Chesapeake Bay Program Office, Annapolis, MD, www.chesapeakebay.net/pubs/ waterqualitycriteria/12022002/Criteria_Final.pdf, accessed 3 August 2007

USEPA (2009) *Chesapeake Bay Phase 5 Community Watershed Model,* In preparation, EPA XXX-X-XX-008 US Environmental Protection Agency, Chesapeake Bay Program Office, Annapolis, MD, January 2009

Wang, H. V. and Johnson, B. J. (2000) 'Validation and application of the second generation three dimensional hydrodynamic model of Chesapeake Bay', *Water Quality and Ecosystem Modeling,* vol 1, no 1–4, pp51–90

Wetzel, R. L. and Neckles, H. A. (1986) 'A model of *Zostera marina* L. photosynthesis and growth: Simulated effects of selected physical-chemical variables and biological interactions', *Aquatic Botany,* vol 26, no 3–4, pp307–323

4

Sustainable Water Management and Non-point Source Pollution Control in Spain and the European Union

Jose Albiac, Mithat Mema and Elena Calvo

Introduction

Water resources have been an important issue in Spain since ancient times. An example is the Contrebia Belaisca bronze from 89BC, which documents the fight between the Iberic town of Salduie (Saragossa) and the Vasconian town of Alaun for the water supply to Salduie (Fatás and Beltrán, 1997); other examples are the dams of Almonacid de la Cuba, Proserpina and Cornalvo, which were the highest built in the whole Roman Empire (Arenillas, 2002). Substantial irrigation projects were undertaken in the Middle Ages under Islamic rule, and the water court of Valencia (Tribunal de las Aguas) is an institution that has been settling water disputes for the past 1000 years.

During the last century, the economic development of agrarian Spain was supported by a succession of hydrological planning and management efforts. The major initiatives were the creation of the water basin authorities in the 1920s, the waterworks plan of the Spanish Republic completed during the dictatorship and the National Hydrological Plans of 1993 and 2001.

These two last plans were a response to the enormous pressure on and degradation of water resources in south-eastern Spain driven by a massive expansion of irrigation from overdrafted aquifers and rivers. The response consisted of large water transfer projects of $4000 hm^3$ in the 1993 plan, and $1200 hm^3$ in the 2001 plan.[1] Both plans experienced strong opposition from political, social and environmental organizations, which finally led to their collapse. The main criticisms were that the plans were based on the traditional supply policy of hydrological planning that was already exhausted, when in fact what were needed were new management initiatives. New initiatives require a mix of measures such as limits on surface and subsurface extractions, protection of the quantity and quality of the river flows and their aquatic ecosystems, concession revision, desalination as a new supply technology, reutilization and regeneration, and water pricing and water markets.

The current water policy in Spain is driven by the National Hydrological Plan of 2005 and the National Irrigation Plan of 2002. The National Hydrological Plan of 2005 is a modification of the 2001 plan, whereby large inter-basin water transfers were stopped in favour of building desalination plants under the AGUA project. However, the AGUA project continues the traditional supply policy, and pretends to solve the scarcity and degradation of water resources in south-eastern Spain by building $600hm^3$ per year of seawater desalination capacity.

The challenges to be confronted in Spain to move towards sustainable management of water resources are considerable, both in terms of water quantity and quality. This chapter describes the situation of water resources in Spain within the European Water Framework Directive context and identifies the main problems that need to be addressed. Spanish and European water policies currently in force are discussed and their shortcomings and achievements evaluated. Finally, the conclusions and policy implications are presented.

Water use by sector, and scarcity and degradation problems

Surface, subsurface and coastal waters have different uses, including domestic, industrial, agricultural irrigation, recreation and support of aquatic ecosystems. Water resources extraction and utilization by sector in Spain in 2002 are presented in Table 4.1. Extractions are close to $40,000hm^3$, of which $6200hm^3$ are used for cooling in electricity production, and $32,000hm^3$ cover the demand from irrigation, water supply companies and other industrial and service sectors. Losses in primary and secondary distribution networks are large and reach $5500hm^3$. Household demand is $2600hm^3$ with an average price of €$1/m^3$, and industrial and service demand is $3200hm^3$ with an average price of €$0.25/m^3$. Net irrigation demand is $20,700hm^3$ and prices are related to the type of agriculture. In inland irrigation areas with collective systems of dams and canals, and field crops of low profitability, prices are below €$0.06/m^3$. In the irrigation areas of eastern and south-eastern Spain with individual pumping from aquifers and high profit crops, the range of prices is between €$0.09/m^3$ and €$0.21/m^3$ (Martínez and Hernández, 2003; Albiac et al, 2006a; INE, 2006).

Table 4.1 *Water resources extraction and utilization by sector in 2002 (hm³)*

	Total	Agriculture	Water companies	Other sectors	Cooling
Extractions	38,200	25,200	5400	1400	6200
Surface	32,500	20,900	4200	1200	6200
Groundwater	5700	4300	1200	200	
Network losses	5500	4500	1000		
Utilization				1400	6200
Agriculture	20,700	20,700			
Households	2600		2600		
Other sectors	3200		1800		
Cooling	6200				

Note: Figures do not include hydropower extractions, estimated at an average of 50,000hm³.
Source: INE (2006) and Martínez and Hernández (2003).

The growing pressure of these economic activities has created problems of water scarcity and quality degradation, mostly linked to groundwater. The more severe problems are found in south-eastern Spain, with pressures coming from agriculture, urban sprawling and tourism on the Mediterranean coast. In inland Spain, surface water resources are under the effective control of basin authorities that manage resources wisely.

Water for cooling and electricity production returns to watercourses and can be used several times with only a small deterioration in quality. However, most water is used for agricultural, urban and industrial purposes, which degrade the quality of return flows. These consumptive uses generate water stress and problems of point and non-point source pollution in watercourses.

The use of water for cooling and electricity production may be cut by half in the coming decades, as a result of more efficient refrigeration systems in power-generating facilities.[2] Agriculture accounts for almost 80 per cent of consumptive water extractions, and the volume of irrigation water could grow to compensate for the effects of climate change in Spain. Urban demand represents 8 per cent of consumptive extractions, and its evolution will be stable since it depends on countervailing factors such as household type, water pricing and technological change that improve water use efficiency.

The effects of climate change in Spain will be severe, according to the IPCC's Fourth Assessment Report. Under scenario A1B in horizon 2100 (Summary for Policy Makers WG1, IPCC, 2007), the main findings are:

- a reduction in precipitations of more than 20 per cent, with a more intense frequency and severity of extreme events such as droughts and floods;
- a sharp drop in water availability which could reach up to 50 per cent in the more arid zones;
- a significant degradation of water quality; and
- diminishing reserves in dams and aquifers.

Irrigation water demand will increase because of the expected reduction of more than 20 per cent in precipitations and higher evapotranspiration values that could be in a range close to 4 per cent.[3] Serious problems of water scarcity are already occurring in the arid and semi-arid regions of Spain, located in the southern and eastern parts of the Iberian Peninsula. The use of irrigation water is very large in these regions, and scarcity problems will worsen because of the large fall of river flows, the increase in irrigation requirements driven by the fall of precipitations, and the increase in water demand for tourism and residential activities in Mediterranean coastal zones. In the coming decades, the effects of climate change could be addressed only by moving towards more sustainable management of water resources.

Human activities linked to water and land resources generate wealth, but these activities also contribute to the degradation of water quality through point and non-point source pollution. To cope with this water degradation, different quality standards have been implemented depending on the final use given to the water. Two alternatives exist to reach the appropriate quality standard: reduction of pollution loads in watercourses, or treatment of the waters being disposed of.

Regarding point source pollution from urban centres and industries, the effects of discharge of residual waters depend on the sewage network and treatment facilities, the industrial production processes, and the type of products consumed by households. In recent decades, there has been a surge in the urban population linked to sewage networks and treatment facilities. An important factor has been the Urban Wastewater Treatment Directive, passed in 1991 and modified in 1998, which requires building secondary treatment plants in urban centres. Spain and other countries in southern Europe, together with France, Belgium and the UK, only have wastewater treatment plants with secondary treatment. Central and northern European countries already have depuration plants with tertiary treatment.[4] The Urban Wastewater Treatment Directive has contributed to a significant reduction of polluting emissions on surface waters, avoiding the subsequent environmental damage to aquatic ecosystems. However, the level of emissions from treatment plants remains high and may cause eutrophication and other problems.

The number of dangerous substances that may affect water quality is high, with very different sources. The manufacturing industry is responsible for most of the emissions of heavy metals (lead, mercury, cadmium), while other substances such as nutrients and pesticides come from agriculture. A few substances have been regulated in the past decades resulting in a fall in their emission, but the emissions abatement is not general. Table 4.2 shows pollutant concentrations in selected Spanish and European rivers. There are important pollution loads by nutrients (nitrates and phosphorus) in the following rivers – Thames (UK), Guadalquivir (Spain), Seine (France) and Escaut (Belgium) – and a high concentration of heavy metals in the Seine (France), Escaut (Belgium), Tajo (Spain), Guadalquivir (Spain) and Porsuk (Turkey).

Table 4.2 *Water quality in selected European rivers (average 2002–2004)*

Country	Watershed	BOD (mg O$_2$/L)	Nitrates (mg N/L)	Phosphorus (mg P/L)	Lead (µg/L)	Cadmium (µg/L)	Chromium (µg/L)	Copper (µg/L)
Norway	Skienselva	2.0*	0.2	0.01	0.2	0.02	0.11	0.62
Sweden	Dalalven	0.1	0.02	0.5*	0.02	0.37*	1.48	
Denmark	Gudena	1.9	1.3	0.09				
UK	Thames	3.4	6.6	0.66	2.9	0.10	1.17	6.63*
Netherlands	Maas	2.5	3.6	0.21	2.8	0.15	1.77	3.77
Belgium	Escaut	3.6	4.7	0.66	12.0	0.67	9.93	10.10
Germany	Rhein	3.0	2.5	0.14	3.0	0.20	2.55	6.22
	Elbe	6.9	3.0	0.17	2.2	0.18	1.20	4.36
	Weser	2.8	3.7	0.14	4.5*	0.20	2.03*	3.56
France	Loire	3.2	3.1	0.21		0.40*		
	Seine	3.1*	5.6	0.63*	22.1*	2.18*	24.67*	15.03*
Spain	Guadalquivir	4.2*	6.1*	0.95*	10.2*	1.87*		5.73*
	Ebro	1.9	2.2	0.09	7.5	0.23*	0.92*	1.61*
	Guadiana	1.6	1.8	0.69*		3.39		
Portugal	Tejo	2.3	1.0	0.20	11.0	3.00	22.33*	2.10
Italy	Po	1.3	2.5	0.25				
Greece	Strimonas		1.8	0.14		0.64*		
Turkey	Porsuk	1.4	1.5	0.06	12.2	6.50	7.50	5.67

Note: The symbol * indicates that the average is for years 1999–2001 or before. The biochemical oxygen demand (BOD) measures pollution by organic matter and water is considered drinkable with a BOD between 0.75 and 1.50 mg O$_2$/L.
Source: OECD (2007).

There has been a reduction of phosphates in detergents used by households, with a fall in the phosphorus load in treatment facilities. Meanwhile, the nitrogen loads from households remain constant. The phosphorus loads received by watercourses originate from urban and industrial point sources and agricultural and livestock non-point sources; most of the nitrogen loads come from non-point agricultural and livestock sources.

Although information on the status of aquatic ecosystems in both Spain and the rest of Europe is scarce, it seems that the improvement in water quality is very slow and in some rivers there is even a worsening in various water quality parameters. This expected improvement should have resulted from the abatement of emissions of organic matter and phosphorus linked to new treatment facilities in urban centres, and the abatement of emissions of heavy metals and chemical substances used by industries.

The data series on water quality in rivers from OECD (2007) show this poor quality improvement that has hampered the recovery of water quality in the past 30 years. The biochemical oxygen demand (BOD) has improved in most European countries except in Belgium (Escaut), the UK (Thames) and The Netherlands (Maas) which show no improvement. An improvement in BOD took place in Germany and Denmark at the beginning of the 1990s, and in France, Spain and Italy at the beginning of the 2000s.

The worst water quality results are for nitrates, with most countries showing no improvement in the past 30 years, and some rivers such as the Loire, Guadalquivir and Strimonas even increasing nitrate loads at the beginning of the 2000s. The only countries that managed to reduce their nitrate loads are Germany (Rhein, Elbe, Wesser) and Norway (Skienselva) during the late 1990s. Phosphorus pollution loads show no improvement in the majority of rivers, with pollution reductions taking place at end of the 1990s in the three rivers in Germany, the Thames (UK), Gudena (Denmark), Maas (The Netherlands) and Ebro (Spain).

Regarding emissions of heavy metals, comparisons are more difficult to make because some countries do not provide data (Denmark, France, Italy and Greece) or the data are not updated (Spain and France). The available data show for the major part abatement of lead, cadmium, chromium and copper pollution loads, with reductions taking place during the middle or late 1990s. Germany shows consistent reductions in the concentration of heavy metals in its rivers, and is the only country with significant improvements in all water quality parameters – BOD, nutrients and heavy metals.

The moderate or nil decrease in pollution loads in all countries except Germany is difficult to understand given the enormous investments in urban treatment plants which were driven by the Urban Wastewater Treatment Directive of 1991. The investments in urban treatment plants in Spain between 1995 and 2005 were €12 billion (Plan Nacional de Saneamiento y Depuración, approved in 1995) and all other European countries have made similar investments with total investments for the EU-15 estimated at €150 billion.

One partial explanation could be that the nitrogen and phosphorus loads coming from agricultural non-point source pollution are not controlled, and these loads may be counterbalancing the abatement gains from urban treatment plants. Another factor could be the increase in non-point source pollution loads from the sprawling of new urbanized areas. In any case, the relative importance of agricultural pollution is increasing, and it seems that between 50 and 90 per cent of the nitrogen loads in surface waters comes from agriculture (EEA, 2005). Pollution problems from agricultural sources are characterized by the uncertainty of the source location, and by the impossibility (or very high cost) of measuring the emission loads of individual farmers. This question has important implications for the design of pollution abatement measures, since non-point source pollution control measures are difficult to apply, and more sophisticated measures are required.

The intensive use of fertilizers is a more severe problem in central and northern European countries than in southern European countries such as Spain. Fertilizer consumption in central and northern countries is above 150kg/ha, while consumption in southern countries is below 150kg/ha.[5] Fertilizer consumption is above 200kg/ha in Germany, Belgium, France, The Netherlands, Ireland and the UK. For example, the nitrogen surplus in soils is

215kg/ha in The Netherlands and 100kg/ha in Belgium and Germany, compared with 40kg/ha in Spain (EEA, 2003), and this surplus is the origin of the nitrate pollution of water bodies. Therefore, the problems of water quality from agricultural non-point source pollution are more serious in central and northern European countries, while the main problem in southern countries such as Spain is water scarcity.

Concern about water scarcity and quality has resulted in the development of an extensive body of rules and regulations in the European Union: the Water Framework Directive (WFD) (2000) and the directives of Drinking Water (1998), Integrated Pollution Prevention and Control (1996), Urban Wastewater Treatment (1991), Nitrates (1991), Dangerous Substances (1976, integrated in WFD in 2006) and Bathing Water Quality (2006).

This legislation has contributed to the abatement of point source pollution from urban and industrial sources, due to the construction of treatment facilities, and the decline in some emissions of dangerous substances from industrial processes. But, as indicated above, the improvement of water quality in European rivers is far from obvious for the majority of basins and pollutants, despite all legislation and investments. There has been a certain improvement of some quality parameters in several surface and coastal water bodies, with the resulting reduction in pressure on their aquatic ecosystems. However, no substantial improvement in the water quality of European rivers is detected, except in the case of Germany. The problems of agricultural non-point source pollution remain, in particular those of nutrients and pesticides (European Commission, 2002), and also the problems of water scarcity in Mediterranean countries.

The Water Framework Directive and water policies in Spain

The European Water Framework Directive is the main legislation initiative to protect water resources and achieve 'good ecological status' for all water bodies. The Directive introduces the principle that water prices should be close to full recovery costs, to improve efficiency in the use of water. Costs must include abstraction, distribution and treatment costs, and also environmental costs and resource value. The Directive establishes a combination of emission limits and water quality standards, with deadlines to achieve appropriate quality for all waters. Water management should be based on basin districts and stakeholder participation, and water pricing at full recovery costs.

European countries defined the basin districts and basin authorities in 2003, and completed the characterization of pressures, impacts and economic analysis of basins in 2004. The results have been used to evaluate the impact of human activities and to identify the areas requiring special protection, guiding the elaboration of the basin management plans and the programmes of measures by

2009. Water pricing policies should be introduced in 2010, and the programmes of measures will be operational in 2012, in order to reach the environmental objectives in 2015. The main scarcity and water quality problems have to be solved in 2021 when the first cycle of management ends and, by 2027, good ecological status of all water bodies has to be achieved.

The principle of cost recovery is one of the key elements in the economic analysis advocated by the Directive. The increase in water prices up to recovery costs is a very interesting measure in the industrial and urban sectors, since there is a demand response to water prices in the industrial and urban water sectors, and higher efficiency in water use is obtained. But in contrast, water demand in irrigation does not respond to water pricing and this fact questions full recovery costs in irrigated agriculture as a valid alternative for water quantity assignment.

Setting some minimum price levels for irrigation water will make farmers understand that water is not a free good. However, using water pricing as a mechanism to allocate water in irrigation is questionable, and Bosworth et al (2002) and Cornish and Perry (2003) show results from the literature and from empirical studies that demonstrate the impossibility of using water prices to assign water in irrigation, both in developed and developing countries. As an alternative to water pricing, these authors indicate that introducing water markets is much more reasonable, although difficult to implement. Therefore, the emphasis of the Directive on water prices is not effective in reducing irrigation demand in Mediterranean countries.

In order to reach the objectives of the WFD, the measure of choice for water scarcity caused by urban and industrial demand is water pricing. Collective irrigation systems based on dams and canal networks should be controlled through command-and-control measures, while irrigation districts based on individual pumping from aquifers need sophisticated incentive schemes that entice the cooperation of farmers in water conservation.

There are some important methodological and information problems within the policy analysis of the WFD, since many basic concepts of environmental policy analysis are not well understood.[6] The emphasis of the WFD on water pricing to achieve water use efficiency and protect the resource follows the Dublin declaration of 1992, but it is a flawed approach. The problem with this 'economic good' approach assumed by the WFD and by many environmental consultants and decision makers in Europe is that the price mechanism can work only where water is a private good (rivalry in consumption and exclusion) which is traded in markets.

Domestic and industrial uses have the characteristics of a private good, but irrigation is different because it has the characteristics of an impure good and environmental externalities. Water pricing could modify consumption where markets exist, such as in urban networks for domestic and industrial demand, but not in agricultural or environmental uses. Furthermore, water markets do not internalize environmental externalities, as seems to be the case in California and Australia. Protection and conservation of water resources, which are common pool resources, require cooperation by the agents managing the resource to achieve collective action.

Another difficulty is the lack of basic statistical information and knowledge of biophysical processes, which favours strategic behaviours by countries, basins and stakeholders in the whole implementation process of the Directive.

The description of the basic measures and supplementary measures in the Water Framework Directive does not make much sense.[7] The writing of measures listed by the WFD does not take into account the state of knowledge in policy analysis from the field of environmental economics. The Directive does not consider either the concepts of private good, public good or externality impacts, and therefore ignores the fact that different types of measures are needed for the different kinds of problems in water resources. The conceptual and empirical misunderstanding in the policy analysis of the Directive is such that there is much confusion among the key water consultants and environment decision makers.

To improve social welfare with policy measures, knowledge is required of biophysical processes, the social benefits and cost functions for each measure, and the optimum social welfare derived from these functions. Without knowing benefit and cost functions for each measure, decision makers can only base policies on cost efficiency. But the WATECO (Working Group on Water Economics of the Common Implementation Strategy) committee in charge of the WFD economic analysis has decided to ignore all this and take as environmental costs whatever environmental expenses countries decide to make.

In order to elaborate reasonable measures, it is essential to clarify the conceptual methodology of policy analysis, and determine the requirements regarding water statistics and scientific knowledge of biophysical processes for the design of measures. Once the biophysical knowledge is generated, the management of water resources is still quite a challenge, because of the public good and environmental externalities aspects of water. The incentives from policy measures should address the strategic behaviour of stakeholders, in order to give rise to cooperation and collective action in conserving water resources.

Another issue is that water institutions are quite frail or non-existent in most European countries, and these countries do not have experience in collecting data, and designing and implementing reasonable water policies that work. For example, Germany and Italy do not have water authorities at either basin or federal levels. In the case of Germany, each state has its own methods, databases and assessment approaches, and achieving policy coordination at basin and federal levels remains to be seen. In the case of Italy, there is a plethora of very small organizations in charge of water, which leads to a serious shortfall in information on water resources and also to difficulties for policy design and implementation.

The European Water Framework Directive approved in 2000 was enacted in Spanish legislation in 2003, just after approval of the Spanish National Hydrological Plan (2001, modified in 2005) and the National Irrigation Plan (2002). The National Hydrological Plan involves large investments (€19 billion) aimed at increasing water supply for agricultural, urban and industrial users. As indicated above, its main project was the Ebro inter-basin transfer to alleviate the severe degradation of water resources in south-eastern Spain, but the transfer was cancelled and substituted by seawater desalination (Albiac et al, 2006b). The National Irrigation Plan involves investments (€6 billion, including the recent Irrigation Crash Plan of 2006) to modernize the largely outdated irrigation facilities, in order to save resources, enhance competitiveness and reduce pollution (MAPA, 2001). Another important piece of legislation in Spain is the Water Quality National Plan of 2007, with investments amounting to €20 billion during the period 2008 to 2015. The objectives of this plan are to upgrade urban treatment plants from secondary to tertiary treatment, build storm tanks in urban centres and protect the sources of water provision.

The National Irrigation Plan promotes investments in modernization through public subsidies. These technological innovations facilitate major reductions in non-point source pollution and increase water conservation because of the higher efficiency of irrigation systems. By contrast, the emphasis of the WFD on source pollution limits, ambient standards and rising water prices is almost irrelevant in irrigation at present. On the one hand, control of irrigation emissions at the source or at the ambient is very difficult to implement and, in addition, there is no biophysical information available to support the design of reasonable measures in the short term. But on the other hand, although water pricing is a good instrument to curb network-connected industrial and urban demand, it is clearly inadequate in terms of allocating water in agriculture and quite inefficient in abating non-point source pollution.[8] The reasons for this are:

- the rigidity of irrigation water demand; and
- the insignificant reduction of pollution achieved by raising water prices, resulting in large losses in income for farmers.

The investments in advanced irrigation technologies currently undertaken by the Spanish administration are much more interesting than the WFD approach, based on pollution limits and water pricing. Updating irrigation technologies does not guarantee the solution to all problems, since the high investment costs may induce dedication to more input-intensive and profitable crops. This more intensive production could eventually increase water demand and pollution loads, but it is obvious that technical innovations in irrigation systems facilitate the private and public control of water quantity and quality.

Thus, it is essential to search for policies based not on a unique measure, but on a combination of institutional, economic and command-and-control

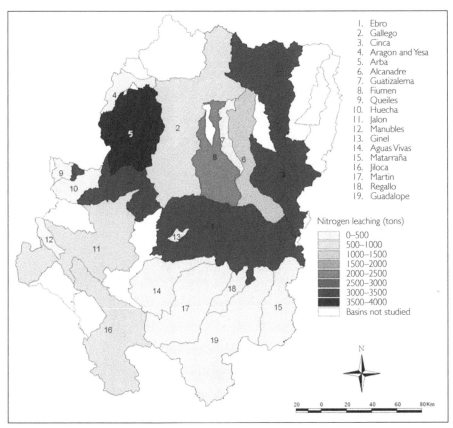

Note: See Plate 10 for a colour version.

Figure 4.1 *Yearly nitrogen emission loads in the middle Ebro basin (t N-NO₃⁻)*

instruments. In the area of non-point source pollution abatement, the focus of recent literature is on incentives based on the ambient pollution loads in watercourses, which are measurable, instead of incentives linked to emission loads at the source (plots). These incentives are tax-subsidy mechanisms and group fines linked to an ambient pollution threshold, which have to incorporate also the strategic behaviour of farmers or their response to measures. The current European legislation on agricultural non-point source pollution is the Nitrates Directive passed in 1991 (European Commission, 1991), which is based on providing information to farmers and voluntary compliance. Since there was no pollution abatement, countries have recently been trying to check the amount of fertilizer applied by asking farmers to present a nitrogen balance book of the farm. Compliance is encouraged by random checks on farmers and the reduction of agricultural subsidies for those found to be not complying. The Directive only applies to irrigation acreage over aquifers which are officially declared polluted, and the accomplishments of this legislation are quite questionable.

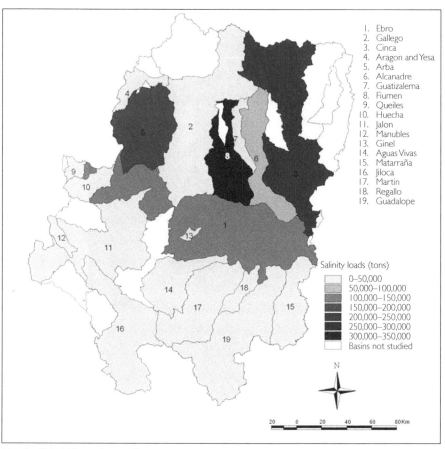

1. Ebro
2. Gallego
3. Cinca
4. Aragon and Yesa
5. Arba
6. Alcanadre
7. Guatizalema
8. Fiumen
9. Queiles
10. Huecha
11. Jalon
12. Manubles
13. Ginel
14. Aguas Vivas
15. Matarraña
16. Jiloca
17. Martin
18. Regallo
19. Guadalope

Salinity loads (tons)

0–50,000
50,000–100,000
100,000–150,000
150,000–200,000
200,000–250,000
250,000–300,000
300,000–350,000
Basins not studied

N

20   0   20   40   60   80 Km

Note: See Plate 11 for a colour version.

Figure 4.2 *Yearly salinity emission loads in the middle Ebro basin (t)*

An example of the large pollution abatement achieved by the Spanish National Irrigation Plan is the study by Mema (2006). The study covers an irrigation acreage of 380,000ha in the middle Ebro basin (the state of Aragon), using a volume of water resources close to 2500hm^3. Irrigation return flows are close to 900hm^3, carrying substantial loads of nitrates and other salts that pollute water-courses. Yearly nitrogen pollution is estimated at around 19,000t N-NO$_3^-$ from nitrogen fertilizers applied that contain 65,000t N-NO$_3$. Salinity pollution is around 1 million tons, mostly from the Flumen, Cinca and Arba basins (Figures 4.1 and 4.2, Plates 10 and 11). The evaluated control measures included taxes and quantitative limits on water use and nitrogen fertilization, taxes on emissions of nitrates and salinity, and investments to modernize irrigation systems.

The results are very relevant for the design of the programme of measures for the WFD. They show that the water pricing instrument advocated by the Water Framework Directive is not adequate to abate nitrogen pollution. Other measures stemming from the WFD are more appropriate in the irrigation context, such as limits on surface and subsurface water extractions, limits on pollution at the source (plots) or standards on ambient pollution (river courses). Irrigation demand does not respond to prices and water pricing is not a cost-efficient measure to abate pollution. Modernizing irrigation systems is a very interesting measure, because it attains the larger pollution abatement with a moderate cost to farmers in terms of quasi-rent. Modernization reduces substantially the use of water input and fertilizer, abates nitrate emission loads by 40 per cent (7000t), and cuts salinity emissions by half (500,000t).

However, the more serious problems of water resources degradation are linked to aquifer overdraft and affect south-eastern Spain. Aquifer overdraft reaches more than 700hm^3 per year in the last decade in the Júcar (160), Segura (220), Sur (70) and Upper Guadiana (220) basins (Figure 4.3, Plate 12). The pressure originates from an intensive agriculture that is very profitable and is based on individual aquifer extractions, together with the pressure of substantial urbanization in the area and tourism on the Mediterranean coast. By contrast, aquifer overdraft in the Upper Guadiana basin is the consequence of a strong expansion of an extensive and low-profit agriculture.

This massive overdraft is the consequence of decades of groundwater mismanagement, despite the fact that groundwater was declared public domain in 1985. Registration of both concessions and private rights of groundwater is far from complete and the number of illegal wells could be above 1 million. The pressure of human activities on aquifers results not only in water scarcity, but also in water quality degradation from pollutants such as salinity and nutrients. The aquifers with higher salinity pollution loads are located in the Sur and Segura basins, and the aquifers with higher nitrogen pollution loads are located in the Guadalquivir and Júcar (Figure 4.3). The negative impact of aquifer overdraft and water quality degradation on aquatic ecosystems is enormous in extended areas of the eastern and south-eastern basins (Hernández et al, 2007).

In contrast, water scarcity and degradation is rather moderate in inland Spain because irrigation is based on collective systems – basin authorities control concessions, river flows and dam reserves, while irrigation user associations manage irrigation districts. The experience and competence of this institutional setting ensures ecological flows, and the management of droughts and floods.

The basin authorities in south-eastern Spain do not control the number of wells or the volume of individual extractions from aquifers linked to very profitable crops, and hence they cannot impose recovery costs. Furthermore, the required price level to curb demand in these areas is above €3/m^3, which is politically unfeasible (Albiac et al, 2006b). In contrast, basin authorities may impose any water price in areas of inland Spain where low-profit crops are grown, because they have absolute control over collective irrigation systems. But the following question then arises: why should they play around allocating water through water pricing when they can make direct and wise water allocations?

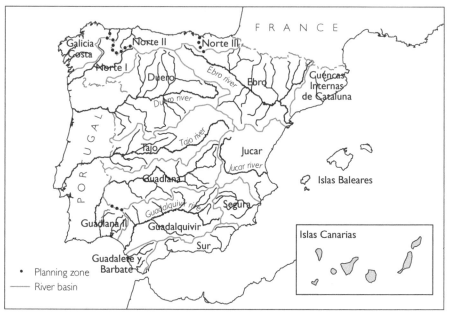

Note: See Plate 12 for a colour version.

Figure 4.3 *River basin authorities in Spain*

There are some examples of unconvincing or clearly erroneous water policies being applied in Spain. Two policies that are most questionable are the Plan of the Upper Guadiana and the AGUA project. The Plan of the Upper Guadiana, recently approved, aims at curbing overdraft in the western La Mancha aquifer and recovering the Tablas de Daimiel natural park, one of the main wetlands in the country. Previous efforts by the basin authority to stop illegal abstractions were turned down by the central Spanish administration, which seems to send the wrong signal not only to those exploiting illegal wells but also to those who use legal wells but pump in excess and deplete the aquifers.[9] Instead of curtailing abstractions, the plan anticipates investments of €5.5 billion to eliminate 220hm³ of overdraft. What is surprising in this enormous investment is that no economic valuation study has been undertaken on the environmental damage caused by the loss of this wetland, to justify the large investment. Furthermore, the large investments in the Upper Guadiana will not work without carefully designed incentives to gain farmers' cooperation. If the plan approach is generalized to the 500hm³ of aquifer overdraft in the Júcar, Segura and Sur basins, then the additional investments needed would amount to €12 billion (Albiac et al, 2007).

The sustainable management of aquifers is not an easy task and it requires cooperation from farmers because they manage the resource and therefore they can take action to conserve the resource. One of the few cases worldwide of collective action in the sustainable management of subsurface water is the eastern

La Mancha aquifer. The aquifer has a surface of $7400km^2$ and the irrigation acreage has expanded from 20,000 to 80,000ha in the past 30 years. The town of Albacete wanted a concession of water for urban use from the Júcar basin authority, and the Júcar basin authority, with support of stakeholders in Valencia state, required the control of extractions in the eastern La Mancha aquifer. Another reason that facilitated the agreement with farmers was the increase in pumping costs because of the overdraft and the fall of the aquifer water table. Farmers in the areas produce extensive crops, which have small profits and high pumping costs as a proportion of total costs, and therefore farmers were willing to cooperate in order to limit extractions and maintain crop profitability.

The support of the Albacete society and farmers led to an agreement between the eastern La Mancha aquifer irrigation association, the state government of Castilla–La Mancha and the Júcar basin authority, to implement sustainable management by achieving collective action and protection of the aquifer. The agreement is based on the inscription of uses previous to 1985 (when subsurface water was not public domain), regularization of uses that started between 1986 and 1997, and a process of characterization of uses and control of pumping extractions. The characterization of uses and control of pumping extractions is made through remote sensing, and information on cultivation plans is provided by each farmer. The key for the system to work is that the farmers are involved in the process of enforcement and control (Ferrer and Gullón, 2004).

The second example of a questionable policy is the AGUA project which was substituted for the cancelled Ebro water transfer to south-eastern Spain. The AGUA project includes investments of €1.2 billion to build desalination plants and expand supply by $600hm^3$, of which $300hm^3$ are for irrigation purposes in the coastal fringe. Although there is a potential irrigation demand in the area from greenhouses and other high profit crops, the pumping costs are much lower than desalination costs, and farmers will not buy desalinated water but rather keep pumping from aquifers. Public investments in desalination are only justified if basin authorities are able to strictly enforce a ban on aquifer overdraft, forcing farmers to buy desalinated water. But the solution found by the water authorities is to subsidize desalinated water up to the level farmers are willing to pay (pumping costs).

An aspect of water management in Spain that should be stressed here is the institutional, technical and organizational competence of basin authorities dating back 100 years. Basin authorities in Spain (Confederaciones Hidrográficas) have a richness of information that is lacking in most European countries, and they are very competent in managing surface water. Each basin authority is made up of watershed boards, where local stakeholders are represented by election. All water management decisions are taken by these watershed boards and therefore stakeholders run the water confederations or Confederaciones Hidrográficas. There is also a high level of competence in the water business sector (construction, distribution, depuration and desalination) and in the dynamic irrigation agriculture of south-eastern Spain.

The problem of achieving sustainable water management in Spain is not a lack of technical capacity, physical capital or human resources, but the absence of political will in the design and implementation of reasonable measures. Solving the degradation and mismanagement of water resources in south-eastern Spain is the key issue for moving towards a sustainable management of water resources in Spain. The example of the eastern La Mancha aquifer presented here shows the type of measures and incentives needed to bring to an end the mismanagement of subsurface water in Júcar, Segura, Sur, Upper Guadiana, and it is also an example for basins worldwide under severe aquifer mismanagement.

Any supply-side policy of expanding water availability, such as the former Ebro inter-basin transfer or the current AGUA project, is questionable as long as groundwater mismanagement continues. Demand-side policies, such as forbidding aquifer overdraft or taxing water abstractions, are technically and politically unfeasible, because basin authorities can only deal at present with surface water. Although there are informal water transactions in south-eastern basins, the introduction of formal water markets requires enormous and persistent efforts. The Water Law was modified in 1999 to promote formal water markets, but it has not spurred any significant transaction in almost ten years. In any case, the introduction of formal water markets would require the control of groundwater. The experience of water markets in Australia and California seems to demonstrate that economic instruments alone fail to protect water resources, and therefore command-and-control and institutional instruments have an important role to play.

The tasks ahead for basin authorities in Spain are quite challenging, since both non-point source pollution and aquifers are common pool resources with impure public good characteristics (public bad in the case of pollution) and with environmental externalities. The sustainable management of water quantity and quality requires that public authorities set up incentives that give rise to cooperation among agents managing the resource, in order to achieve the collective action needed for water conservation.

Conclusions and policy implications

One of the important environmental questions in Spain and the European Union is the scarcity and degradation of water resources. The main pressures in Spain derive from the urban, industrial and irrigation consumptive uses, which create water scarcity and widespread water quality degradation from point and non-point source pollution. Spain is a semi-arid region with a massive use of water for irrigation. The scarcity outlook could deteriorate because of the possible expansion of irrigated acreage and the pressure of urbanization and tourism in coastal areas, and because climate change will reduce available resources.

The efforts to curb pollution were started throughout the whole European Union through several European directives. This legislation addressed the effects of point source pollution emissions from urban and industrial discharges, which depend on sewage collection and treatment facilities. Despite these efforts undertaken by public administrations in past decades, pollution by nutrients and heavy metals remains high in many watersheds of the more important river basins in Spain and the rest of Europe. The extensive European regulation has facilitated large investments in water treatment plants and technological innovations in industries and households, which have limited or reduced the emissions of some pollutants, but the abatement of emissions is not general. The efforts on urban and industrial point source emissions should continue, and effective control on non-point source pollution is needed, such as abatement of nutrients and pesticides from agriculture.

The future of water resources in Spain will depend on the management measures taken. Water scarcity could worsen considerably by further uncontrolled extractions and the effects of climate change. Solving the scarcity problem may require reallocating some water from off-stream use by agricultural, urban and industrial users to environmental uses both in aquifers and streams, and also in the coastal wetlands. There are serious problems of water quality degradation not only in Spain but in almost all European countries.

The case of Spain shows that the implementation of the Water Framework Directive is not an easy task. Both the Spanish Ministry of Environment and the European Commission Environment Directorate advocate water pricing in irrigation and using the Common Agricultural Policy (CAP) to penalize farmers. Research projects funded by the European Commission and some other studies also recommend these flawed policy options.[10]

But the problems of scarcity and quality degradation cannot be solved with these two policies. Water pricing is a very good instrument for industrial and domestic demand, but it is ineffective for irrigation. Water pricing is not a workable option because:

- there is no control over the huge number of illegal wells and the quantities pumped from aquifers;
- water shadow prices are above €3/m^3, a price that is politically unfeasible since desalination costs are €0.50/m^3 and urban water prices are around €1/m^3; and
- the administration lacks the information on aquifer dynamics precluding the enforcement of sustainable extractions.

The CAP is also ineffective in influencing water extractions in south-eastern Spain, because CAP subsidies are targeted towards continental products such as field crops, whereas production in the area consists of Mediterranean crops such as fruits and vegetables which have negligible CAP subsidies.

The investments in advanced irrigation technologies currently undertaken by the Spanish administration are much more interesting than the WFD approach, based on pollution limits and water pricing. The Spanish National Irrigation Plan has an important potential for large water savings and pollution abatement, through the investments in irrigation technologies. These technologies facilitate the private and public control over water quantity and quality, and their potential depends on the right coordination and collaboration between farmers, their water user associations and water authorities. The example given of the effects of irrigation modernization in the middle Ebro basin, indicate that nitrogen and salinity non-point source pollution is cut by half when investing in advanced irrigation technologies.

While irrigation modernization undertaken by Spanish administration is a very good measure, other water policy measures such as the AGUA project and the Special Plan of the Upper Guadiana seem rather questionable.

The design and implementation of reasonable measures required by the WFD is a difficult task not only in Spain, but also in the whole European Union. The improvement in the management of water resources requires better information and knowledge on surface and subsurface resources and their associated ecosystems. These tasks need time and resources because of the complex biophysical, spatial and dymanic dimensions involved. At present, data on water quantity are not very good in the European Union, and data on water quality are even more limited. For example, the quantity figures of the European Environment Agency do not match national figures in Spain and France, and water quantity information from countries such as Italy is not available.

The policy analysis of the WFD needs substantial improvement in both the methodological approach and the choice of instruments. The 'economic good' perspective that follows the Dublin declaration is flawed, because the price mechanism can work only when water is a private good. This may be the case in urban networks for domestic and industrial demand, but not for agriculture or environmental uses. Additionally, water markets cannot internalize environmental externalities. The common pool characteristics and environmental externalities of water resources call for cooperation and collective action by stakeholders, and not for economic instruments.

The decision by the WATECO committee (in charge of the WFD economic analysis) to take as environmental costs whatever environmental expenses countries decide to make, highlights the weaknesses and drawbacks of the current water policy analysis in Europe. WATECO ignores both the principle of welfare optimization derived from the benefit and cost functions of measures, but also the principle of cost efficiency used when the benefit function is unknown.

Knowledge of the underlying biophysical processes is critical for water management, especially for managing aquifers and controlling non-point source pollution, and this requires the availability of basic facts on aquifer and pollution characteristics and dynamics at local watershed scale. Regarding pollution, information is needed on the emission loads, the pollutants' transport and fate processes, and the ambient pollution in watercourses. Also, the lack of economic valuation of damage to aquatic ecosystems from aquifer overdraft and non-point source pollution, precludes the assessment of the benefits of policy measures.

Even when all the biophysical knowledge is available, managing the quantity and quality of surface and groundwater is quite challenging because of the public good characteristics of water and the associated environmental externalities. The design of measures must take into account the strategic behaviour of water stakeholders, setting up incentives for cooperation in order to achieve water conservation through their collective action. Both aspects – biophysical knowledge and collective action – are unlikely to be in place in any European (or non-European) country by 2015, when the 'good ecological status' objective of the WFD is supposed to be attained.

Most European countries have no experience in collecting data to design and implement reasonable water policies, because their water institutions are frail or non-existent. Two examples are Germany and Italy, which do not have water authorities at basin or central government level. The potential for change in European water policies points towards mild improvements in coming decades, when the required institutional setting and collective action by stakeholders could be progressively achieved, but then the climate change impacts would be a real challenge calling for a quantum leap improvement towards sustainable water management.

Acknowledgements

Support for the research results presented here were provided by projects CICYT AGL200508020C05 and AGL200765548C02, and INIA RTA04-141-C2 from the Spanish Ministry of Education and Science; and project INCO CT2005015031 from the Directorate of Research of the European Commission. The Spanish Ministry of Foreign Affairs (AECI) and the University of Zaragoza provided funds to support the doctoral programme of Mithat Mema.

Notes

1   1 cubic hectometre ($1hm^3$) equals 1 million cubic metres ($10^6 m^3$). The investments of the 1993 plan amounted to €28 billion (6 per cent of GDP) and those of the 2001 plan amounted to €19 billion (3 per cent of GDP).

2   The new tower cooling systems reduce the amount of water by two orders of magnitude per megawatt-hour, compared with current refrigeration systems with single circulation.

3   No estimate is available for Spain. The figure of 4 per cent is the range considered by CSIRO (2007) for Australia as the annual increase of evapotranspiration in horizon 2100.

4   Tertiary treatment is more advanced than secondary treatment and reduces the emission loads of the nutrients phosphorus (up to 60 per cent) and nitrogen (up to 90 per cent).

5   Fertilizer consumption corresponds to the sum of nitrogen (N), phosphorus (P_2O_5) and potassium (K_2O).

6   Such as objectives, instruments (institutional, economic, command-and-control), welfare optimum, target, cost efficiency, private good, common pool resource, stakeholders' cooperation and collective action.

7   The definition of 'basic measures' in the WFD shows that they are not policy measures at all, but a reformulation of the objectives that are supposed to be reached with previous water legislation. The definition of 'supplementary measures' in the WFD is overly general and does not have any practical application.

8   Martínez and Albiac (2004 and 2006) prove the inefficiency of water pricing to abate nitrate pollution. See also Cornish et al (2004) on water pricing, summarizing results from Bosworth et al (2002) and Cornish and Perry (2003).

9   In 2005, the Guadiana basin authority documented 5000 illegal wells and sent the cases to court. Then, the federal Ministry of Environment fired the president and the water commissioner of the Guadiana basin authority.

10   An example is Downward and Taylor (2007) on Almería, which states that sustainable management can be achieved by water pricing and augmenting water supply through desalination. Irrigation water use in Almería is around 260hm³ and domestic and industrial use is around 90hm³. Water pricing may reduce industrial and domestic demand, but not irrigation aquifer pumping. Since the growing urbanization pressure on the coast will take over any water pricing savings in industry and urban demand, scarcity from irrigation aquifer overdraft will continue. Desalination cannot work either, because farmers will not buy desalinated water unless a strict enforcement of overdraft is in place, a daunting task for authorities. The implication is that the measures advocated by Downward and Taylor cannot deliver the collective action required for water conservation. Several examples from EU research projects advising questionable water policies are the following: WFD meets CAP (www.ecologic.de/modules.php?name=News&file=article&sid=1369), Aquamoney (www.aquamoney.org), AquaStress (www.aquastress.net), WADI (www.uco.es/investiga/grupos/wadi), POPA-CTDA (www.popa-ctda.net) and POLAGWAT (http://susproc.jrc.es/docs/waterdocs/FinalRep150802.pdf).

References

Albiac, J., Hanemann, M., Calatrava, J., Uche, J. and Tapia, J. (2006a) 'The rise and fall of the Ebro water transfer', *Natural Resources Journal,* vol 3, no 46, pp727–757

Albiac, J., Martínez, Y. and Tapia, J. (2006b) 'Water quantity and quality issues in Mediterranean agriculture,' in OECD (eds) *Water and Agriculture: Sustainability, Markets and Policies,* OECD, Paris

Albiac, J., Martínez, Y. and Xabadía, A. (2007) 'El desafío de la gestión de los recursos hídricos', *Papeles de Economía Española,* no 113, pp96–107

Arenillas, M. (2002) 'Obras hidraúlicas romanas en Hispania', http://traianus.rediris.es/textos/hidraulicas.htm, accessed 15 June 2008

Bosworth, B., Cornish, G., Perry, C. and Van Steenbergen, F. (2002) *Water Charging in Irrigated Agriculture: Lessons from the Literature*, Report OD 145, HR Wallingford, Wallingford

Cornish, G. and Perry, C. (2003) *Water Charging in Irrigated Agriculture: Lessons from the Field*, Report OD 150, HR Wallingford, Wallingford

Cornish, G., Bosworth, B., Perry, C. and Burke, J. (2004) *Water Charging in Irrigated Agriculture: An Analysis of International Experience*, FAO Water reports No 28, FAO, Rome

CSIRO (2007) *Climate Change in Australia. Technical Report 2007*, CSIRO, Clayton South

Downward, S. and Taylor, R. (2007) 'An assessment of Spain's Programa AGUA and its implications for sustainable water management in the province of Almería, southeast Spain', *Journal of Environmental Management,* no 82, pp277–289

EEA (European Environment Agency) (2003) *Europe's Water: An Indicator-based Assessment*, Topic Report No 1, EEA, Copenhagen

EEA (2005) *European Environmental Outlook*, EEA Report No 4, EEA, Copenhagen

European Commission (1991) *Concerning the Protection of Waters against Pollution caused by Nitrates from Agricultural Sources*, Council Directive 91/676/EEC (Nitrates Directive), Office for Official Publications of the European Communities, Luxembourg

European Commission (2002) *Implementation of Council Directive 91/676/EEC Concerning the Protection of Waters against Pollution caused by Nitrates from Agricultural Sources. Synthesis from Year 2000 Member States Reports*, Report COM(2002)407, Directorate-General for Environment, Office for Official Publications of the European Communities, Luxembourg

Fatás, G. and Beltrán, M. (1997) 'Salduie, ciudad ibérica' in *Historia de Zaragoza* vol 1, Ayuntamiento de Zaragoza-Caja de Ahorros de la Inmaculada, Zaragoza

Ferrer, J. and Gullón, N. (2004) 'Actuaciones de gestión y regularización administrativa en el acuífero de Mancha oriental', Paper presented at the VIII Simposio de Hidrogeología de la Asociación Española de Hidrogeología, Zaragoza

Hernández, N., Martínez, L., Llamas, M. and Custodio, E. (2007) *Groundwater Issues in Southern EU Member States: Spain Country Report*, Report presented to the Secretariat of EASAC (European Academies of Sciences Advisory Council), Madrid

INE (2006) *Bases de Datos sobre Estadísticas Medioambientales del Agua* and *Cuentas Satélite del Agua*, Instituto Nacional de Estadística, Madrid, www.ine.es

IPCC (2007) *Climate Change 2007: Synthesis Report. Contribution of Working Groups I, II and III to the Fourth Assessment Report of the Intergovernmental Panel on Climate Change*, Core Writing Team, R. K. Pachauri and A. Reisinger, Intergovernmental Panel on Climate Change, Geneva

MAPA (2001) *El Plan Nacional de Regadíos: Horizonte 2008*, Ministerio de Agricultura, Pesca y Alimentación, Madrid

Martínez, Y. and Albiac, J. (2004) 'Agricultural pollution control under Spanish and European environmental policies', *Water Resources Research*, vol 40, no 10, doi:10.1029/2004WR003102

Martínez, Y. and Albiac, J. (2006) 'Nitrate pollution control under soil heterogeneity', *Land Use Policy*, vol 4, no 23, pp521–532

Martínez, L. and Hernández, N. (2003) 'The role of groundwater in Spain's water policy', *Water Internacional*, vol 3, no 28, pp313–320

Mema, M. (2006) 'Las políticas de control de la contaminación difusa en el Valle medio del Ebro', PhD thesis, University of Zaragoza, Zaragoza

OECD (2007) *OECD Environmental Data. Compendium 2006*, Organisation for Economic Co-operation and Development, Paris

5

Non-point Pollution Regulation Approaches in the US

Marc O. Ribaudo

Introduction

More than 440 million acres in the US (19.5 per cent of land) is dedicated to growing crops, and another 587 million acres (26 per cent) is in pasture and range, largely used for domestic livestock production (Lubowski et al, 2006). Agricultural activities on these lands produce a plentiful, diverse and relatively inexpensive supply of food, feed and fibre for people in the US and abroad. However, agricultural production practices may degrade the environment. Soil erosion, nutrient and pesticide run-off, and irrigation can pollute water resources. The extent and degree of the environmental problems associated with agriculture vary widely across the country. Concern over these problems has given rise to local, state and federal conservation and environmental policies and programmes to address them. However, agricultural pollution is not looked upon in the same way as pollution from sewage treatment plants, factories and other point sources. This is reflected in the approaches used to address agricultural pollution.

While no comprehensive national study of agriculture and water quality has been conducted, the magnitude of the impacts can be inferred from several water quality assessment studies. Based on state assessments, the Environmental Protection Agency (USEPA) concluded in its 2000 Water Quality Inventory that agriculture is the leading source of pollution in 48 per cent of river miles, 41 per cent of lake acres (excluding the Great Lakes) and 18 per cent of estuarine waters found to be water-quality impaired (USEPA, 2002). This makes agriculture the leading source of impairment in the nation's rivers and lakes, and a major source of impairment in estuaries. Agriculture's contribution has remained relatively unchanged over the past decade.

The significance of water pollutants commonly produced by agriculture is suggested by information on impaired waters provided by states, tribes and territories to the USEPA in accordance with Section 303(d) of the Clean Water Act. These are waters that do not meet water quality standards and cannot meet those standards through point-source controls alone. The most recent information indicates that 25,823 bodies of water (stream reaches or lakes) are impaired nationwide (USEPA, 2005). Pathogens, sediment and nutrients are among the top sources of impairment, and agriculture is a major source of these pollutants in many areas.

A US Geological Survey (USGS) study of agricultural land in watersheds with poor water quality estimated that 71 per cent of US cropland (nearly 300 million acres) is located in watersheds where the concentration of at least one of four common surface-water contaminants (nitrate, phosphorus, faecal coliform bacteria and suspended sediment) exceeded generally accepted instream criteria for supporting water-based recreation activities (Smith et al, 1994). Another USGS study found that structural changes in animal agriculture between 1982 and 1997 put upward pressure on stream concentrations of faecal coliform bacteria in many areas of the Great Plains, Ozarks and Carolinas (Smith et al, 2005).

The economic damage from agricultural pollution is largely unknown. Research from the 1980s estimated that soil erosion from cropland was causing between US$5 billion and US$18 billion worth of damage each year (Ribaudo, 1989). No comparable estimates are available for the damage from nutrients and pesticides.

Characteristics of non-point source pollution

Non-point source (NPS) pollution has several important characteristics that influence how different policies for controlling it may perform. NPS pollution loadings depend in part on random variables such as wind, rainfall and temperature, making it a stochastic process. As a result, a particular policy will produce a distribution of water quality outcomes rather than a single outcome (Braden and Segerson, 1993; Ribaudo et al, 1999). This by itself does not prevent attainment of *ex ante* efficiency through the use of standard instruments. However, it implies that a policy must be designed to consider 'moments' or 'points' of the distribution other than the mean. For example, nearly all soil erosion occurs during extremely heavy rain events. Practices that control erosion from 'average' rainfalls but fail under heavy rain events will generally be ineffective in protecting water resources from sediment inflows.

The characteristics of agricultural NPS pollution vary over geographic space, due to the great variety of farming practices, land forms and hydrologic characteristics found across even relatively small areas. An effective policy tool should be flexible enough to work in many different circumstances (Ribaudo et al, 1999).

The most problematic characteristic from a policy standpoint is the inability to observe emissions. NPS pollution enters water systems over a broad front. Changes in ambient water quality can be observed and aggregate loadings of agricultural chemicals and sediment can be estimated, but the sources of these residuals cannot be observed. In addition, monitoring the movement of NPS emissions is often impractical or prohibitively expensive. The inability to observe emissions would not be such an obstacle if there were strong correlations between emissions and some observable aspect of the production process, or between emissions and ambient quality. A policy could then be directed at the production process or at ambient quality. For example, if a shallow aquifer that is entirely overlain by cropland is threatened by agricultural chemicals, then a policy could be targeted at chemical use on that cropland. However, such correlations are unlikely to occur, and where relationships can be established, they are unlikely to be the same across a range of conditions. Thus, from a regulatory perspective, NPS pollution involves a moral hazard in regulation (Malik et al, 1992). Although the regulatory agency can judge the quality of a water body through biological and chemical measurements, the agency cannot determine whether the observed state of water quality is caused by the failure of non-point sources to take appropriate actions or undesirable states of nature (high rainfall, for example).

Furthermore, production inputs critical for predicting or forming expectations on NPS pollution may also be unobservable or prohibitively expensive to monitor. For example, there is a close correlation between chemical contamination of groundwater and the amount of applied chemical and soil type. Chemical characteristics and soil type can be observed, but the amount of a chemical reaching an aquifer also depends on timing and method of application. These activities are generally not observable to a regulating agency without very costly and intrusive monitoring (Segerson, 1999).

US laws addressing non-point source pollution

Responsibility for NPS pollution control was given to the states at the outset of current federal water quality law. Non-point sources of water pollution were first identified as necessary for control in the 1972 amendments to the Federal Water Pollution and Control Act, later known as the Clean Water Act (CWA). However, agricultural stormwater discharges and irrigation return flows were specifically exempted from permit requirements that point sources faced. Instead, Section 208 called for the development and implementation of 'area-wide' water-quality management programmes to ensure adequate control of all sources of pollutants, point and non-point, in areas where water quality was impaired. The CWA directed states to develop plans for reducing NPS pollution, including appropriate land management controls. Congress also provided support for research programmes at the US Department of Agriculture (USDA) to improve run-off management on farms. The 1977 amendments further emphasized the role of NPS pollution control in meeting water quality goals.

The Section 208 process was generally not regarded as a success (Harrington et al, 1985; USEPA, 1988; Cook et al, 1991). A series of House Public Works and Transportation subcommittee hearings found that administration technical and financial support for the programme was lacking, coordination with the point source programme was non-existent, and the data necessary for implementing an effective programme were inadequate (Copeland and Zinn, 1986). The consequence was that states lagged in the development of area-wide management programmes, and the USEPA could not readily judge whether the Section 208 plans finally developed were adequate for achieving NPS goals. The USEPA was also not given effective enforcement tools to ensure that NPS management plans were viable or actually implemented (Wicker, 1979).

Part of the reason for the lack of progress was the relative unimportance given to NPS pollution. Point source pollution was seen as the more serious problem, being responsible for the most visible water quality problems. Point sources were also easier to control through centralized technology standards, in the form of the National Pollutant Discharge Elimination System permit system. As a result, greater effort and resources were devoted to point source pollution, with little dissent from environmental or other groups.

By the late 1980s, the USEPA started taking a harder look at NPS pollution as an important cause of remaining water quality problems. While point source discharges were still causing problems, NPS pollution had become the largest unregulated source of pollution. In its 1984 Report to Congress, the USEPA stated that: 'In many parts of the country, pollutant loads from non-point sources present continuing problems for achieving water quality goals and maintaining designated uses' (page 1-1). The report also singled out agriculture as 'the most pervasive cause of non-point source water quality problems' (page 2-6).

Congress responded by revamping the NPS programme in the Water Quality Act (WQA) of 1987. The WQA placed special emphasis on NPS pollution by amending the Clean Water Act's Declaration of Goals and Policy to focus on the control of non-point sources of pollution (USEPA, 1988). Section 319 of the Act requires each State to:

- identify navigable waters that, without additional action to control non-point sources of pollution, cannot reasonably be expected to attain or maintain applicable water-quality standards or goals;
- identify non-point sources that add significant amounts of pollution to affected water; and
- develop an NPS management plan on a watershed basis to control and reduce specific non-point sources of pollution.

Among other things, the management plan is required to contain a list of best management practices (BMPs) for controlling NPS pollution, a timetable for implementing the plan and enforceable measures to ensure the plan is implemented. NPS control plans can include state regulatory measures, but usually emphasize voluntary actions like those used in USDA conservation programmes. Implementation grants to states and tribes – US$200 million in fiscal year (FY) 2007 – fund projects such as the installation of BMPs for dealing with animal waste; design and implementation of BMP systems for stream, lake and estuary watersheds; and basin-wide landowner education programmes. The Clean Water State Revolving Fund (CWSRF), created by Congress to fund the construction of water treatment plants, can be used by states to provide reduced-rate loans for water quality projects included in the state NPS plan. Fifteen states have used CWSRF for funding waste management systems, manure spreaders, conservation tillage equipment, irrigation equipment, filter strips and streambank stabilization. The Water Quality Act also authorized federal loan and grant funds to help states develop and implement NPS control programmes. All states currently have federally approved NPS management plans.

The decentralized control called for in the Clean Water Act does not easily address the problem of interstate transport of pollutants (transboundary issues). Whereas most of the problems from NPS pollution are felt close to the source, some NPS pollutants can travel long distances in major rivers or affect regional water bodies such as the Gulf of Mexico or Chesapeake Bay. The beneficiaries of a state's pollution control policies could therefore be residents of other states. There are very few examples where states have come together without federal prodding to address regional water quality issues, despite common goals and the fact that an individual state may not be able to meet water quality goals without better control of interstate pollution. Cooperation would increase the likelihood of a more efficient response to pollution problems, in that a greater share of those who benefit is accounted for.

In contrast to most agricultural sources, water pollution from some animal feeding operations is treated as a point source under the Clean Water Act. Confined animal feeding operations that meet certain size thresholds and other conditions fall under the National Pollutant Discharge Elimination System (NPDES). These operations, known as concentrated animal feeding operations (CAFOs), must obtain NPDES permits that specify standards for the production area (i.e., housing, waste storage). CAFOs must also implement a nutrient management plan for animal waste applied to land in order to qualify for the agricultural stormwater exemption.

The USEPA estimates that up to 15,500 operations are covered by the CAFO regulations. These regulations may impose significant manure management costs in areas where land for spreading manure is scarce (Ribaudo et al, 2003). These costs could influence location decisions for large operations and spur the development of alternative uses for manure. The USEPA encourages CAFOs to seek financial and technical assistance from the USDA to help them meet manure management requirements.

Over the past five years, there has been a marked increase in the use of nutrient management plans on animal feeding operations. Part of this is due to USEPA's focus on animal feeding operations. Individual states have also taken action to reduce the water quality impacts of animal operations. Animal feeding operations may also be adopting nutrient management as a means of demonstrating due care to protect themselves from citizen complaints, particularly where communities and farms are in close proximity with each other.

The total maximum daily load (TMDL) provisions of the Clean Water Act are intended to be the second line of defence for protecting the quality of surface-water resources. When technology-based controls are inadequate to ensure that water quality meets state standards, Section 303(d) of the Clean Water Act requires states to develop TMDLs for affected waters. A TMDL is the maximum amount of a pollutant that a water body can receive and still meet water quality standards, and an allocation of that amount to all the pollutant's sources. States must submit to the USEPA a list of impaired waters and the cause of the impairment. More than 20,000 such waters have been identified as impaired under Section 303(d). Although NPS discharges are included in the TMDL, the provisions do not require states to implement regulations to reduce them. However, the states can use regulations to control NPS pollution if necessary. The TMDL provisions were little used until about 15 years ago, when pressure from environmental groups forced the USEPA and states to accelerate the development of TMDLs for impaired waters.

A separate federal NPS pollution control programme was implemented for the coastal zone.[1] The Coastal Zone Management Act Reauthorization Amendments (CZARA) of 1990 added NPS water pollution requirements to the Coastal Zone Management Act of 1972. The Coastal Zone Management Act is a collaboration between federal, state and local governments for managing and allocating coastal resources. All coastal states, including Great Lakes states, can develop a coastal zone management programme and receive financial assistance from the federal government (developing a programme is not mandatory). CZARA requires that each state and territory with an approved coastal zone management programme submit a plan to implement management measures for NPS pollution to restore and protect coastal waters. Currently, 34 coastal states and territories have developed NPS pollution control plans.

The Safe Drinking Water Act (SDWA) of 1974 requires the USEPA to set standards for drinking-water quality and requirements for water treatment by public water systems. States are required to develop Source Water Assessment Programs to assess the areas serving as public sources of drinking water in order to identify potential threats and to initiate protection efforts. The USEPA is required to establish a list of contaminants for consideration in future regulation. The Drinking Water Contaminant Candidate List, released in March 1998, lists several agricultural chemicals – including metolachlor, metribuzin and the triazines – for consideration. The Safe Drinking Water Act also requires farms serving water through pipes or other constructed conveyances to an average of 25 people, or more than 15 service connections, for more than 59 days/year to meet drinking water regulations. The Act prevents farmers from injecting any contaminant into an underground source of drinking water, or using a well if the contaminant may cause a violation of any primary drinking water regulation or adversely affect human health.

The Federal Insecticide, Fungicide and Rodenticide Act (FIFRA) provides direct controls over the sale and use of pesticides. Under FIFRA, all pesticides must be approved by the USEPA through a mandatory registration process. Products determined to pose an unacceptable risk to human health or to the environment, including water quality, can be denied registration, thereby preventing their distribution and use. Fifty pesticides and pesticide formulations have been banned under FIFRA as of 2004.

It should be noted that the USDA plays a major role in addressing NPS pollution in the US. The CWA specifically mentions the USDA as being the primary source of assistance for farmers implementing practices to reduce NPS pollution. The USDA employs a number of programmes to help farmers reduce polluted run-off from farms. The largest is the Environmental Quality Incentive Program (EQIP). EQIP provides financial and technical assistance to farmers for installing management practices that protect water resources. Annual funding is currently about US$1.3 billion. From 1997 through to 2004, 37 per cent of EQIP funds were spent on water quality and water conservation-related practices; another 28 per cent were spent on managing livestock manure nutrients, which is a major source of water pollution.

The Conservation Security Program (CSP) also supports conservation measures that protect water quality. However, it takes a different approach by rewarding good stewards for maintaining 'good' management practices. CSP also encourages increased environmental performance by offering higher incentive rates for additional management measures. CSP's budget is much smaller than EQIP's and it is targeted to specific watersheds. Other programmes, such as the Conservation Reserve Program, Wetland Reserve Program and Farmland Protection Program also provide water quality benefits, even if water quality is not their primary goal.

Federal water quality laws have largely passed on responsibility for developing NPS programmes to the states, and have allowed the states to use the full range of policy tools, including voluntary (education, technical assistance), regulatory (technology and performance standards) and economic incentives (taxes, subsidies, trading) to comply with federal requirements. Early on, states developed programmes almost exclusively around voluntary approaches, supported with some cost sharing. Voluntary approaches have not provided the level of protection that is often required to achieve water quality goals (ELI, 2000). In recent years, more states have developed programmes that contain non-voluntary elements, or enforceable mechanisms.

Enforceable policy instruments for NPS pollution

A number of regulatory approaches have appeared in US water quality programmes at the state and federal level (Table 5.1). The tools that are used and how policies are implemented determine the incentives for adopting environmental quality-enhancing practices. The following section reviews the five general non-voluntary approaches to agricultural non-point source pollution control that are currently being used.

Compliance

A quasi-regulatory approach that is used by the USDA is called conservation compliance. Conservation compliance requires a basic level of environmental compliance as a condition of eligibility for other agriculture programmes. This tool shares characteristics with both government standards for private goods/actions and economic incentives. It is similar to the former in that the government establishes a set of approved practices, except that here compliance is linked to a direct economic payment. Because existing programmes are used for leverage, compliance mechanisms require no budget outlay for producer payments, although considerable technical assistance is needed to develop conservation compliance plans. Compliance mechanisms were enacted primarily as a method for removing apparent inconsistencies between farm income support programmes (which can encourage more intensive agriculture) and conservation programmes. A weakness of compliance is that a reduction in subsidies, due to programme changes or high prices, reduces its effectiveness. Also, the incentive applies only to those producers participating in agricultural programmes.

Evidence suggests that compliance does have an effect. Reductions in excess erosion (i.e., erosion in excess of the sustainable rate) were larger on farms that received farm programme payments than on farms that did not. Excess wind erosion declined by 31 per cent on farms receiving payments, but by only 14 per cent on farms not receiving payments. Excess water erosion dropped by 47 per cent on farms receiving payments and by 41 per cent on farms not receiving payments.

Technology standards

The most common mechanisms employed in regulatory programmes are technology standards. These generally call for farmers to implement a unique conservation plan that contains recommended BMPs. Many states apply this

Table 5.1 *Policy instruments used in the US*

Policy tool	Participation	Government role	Selected US programmes Programme title
Educational/ Technical Assistance Government	Voluntary	Provide farmers with information and training to plan and implement practices	Conservation Technical Assistance
Labelling Standards for Private Goods	Voluntary, but standard must be met for certification	Government sets standards, which must be met for certification, typically involving voluntary 'eco-labelling' guidelines	Organic certification
Incentive Policies: Land Retirement Payments	Voluntary	Annual payments for retiring land from crop production for contract duration; contracts generally long term (10 years – permanent)	Conservation Reserve Program, Wetland Reserve Program and Emergency Wetland Reserve Program
Incentive Policies: Financial Assistance	Voluntary	Payments to offset the cost of adopting specified best management practices. Payments may originate from an environmental credit trading programme	Environmental Quality Incentives Program, Wildlife Habitat Incentives Program, Conservation Security Program, Farm and Ranch Lands Protection Program and Grassland Reserve Program
Incentive Policies: Environmental Taxes	Involuntary, but payment amount depends on behaviour	Per-unit charges for failure to meet environmental goals	None at the Federal level
Compliance Mechanisms	Involuntary, after opt-in to Commodity Programs	Sets standards for environmental performance and determines whether requirements are met before releasing payments	Highly Erodible Land Conservation (Conservation Compliance and Sodbuster) and Wetlands Conservation (Swampbuster)
Regulatory Requirements	Involuntary	Producers subject to regulations if voluntary measures do not achieve environmental goals. Operations may be subject to effluent discharge permits. Use restrictions and bans on certain pesticides. Farmers may not 'take' a member of a listed species; agencies must protect and restore species and their habitats	CZARA, CWA (animal feeding operations), FIFRA and Endangered Species Act

approach either uniformly across the state (non-targeted) or targeted to specific geographic areas. Non-targeted technology standards require farmers to adopt a conservation plan containing management practices generally believed to represent 'good stewardship'. A few states have developed a list of approved BMPs (Kentucky has a list of 58 practices for example), whereas others are less specific up front as to what a plan should contain. All plans must be approved by the state. Laws directed at crop production generally allow voluntary adoption at first, with a regulatory backup. Enforcement is generally through citizen complaint. If a suitable plan had been adopted and in force, the producer would not be subject to fines or penalties if a citizen files a complaint for damages, and may receive state assistance to alter the plan to address the specific complaint.

The Clean Water Act requires that CAFOs have a nutrient management plan. A number of states have similar requirements and include other measures such as setbacks, buffers and restrictions on where animal waste can be applied to land. For example, Oklahoma has banned the application of chicken litter in a watershed where a drinking water reservoir is threatened with pollution.

A desirable characteristic of a regulatory tool is flexibility (Ribaudo et al, 1999). Technology standards cannot be considered flexible, which reduces the economic efficiency of the approach. States can achieve some limited flexibility through administrative means by not setting specific water quality standards or goals, but instead requiring the more vague 'better stewardship'. This leads to the acceptance of a wide range of conservation plans that do not greatly constrain farmers.

The effectiveness of a regulatory approach depends on enforcement. A common enforcement approach by states is to rely on citizen complaint. But a problem with technology standards that rely on citizen complaints for enforcement is that they do not provide adequate incentives for the landowner to implement an efficient amount of pollution control, or for the potential victim to make known the costs of pollution. NPS pollution is characterized by an inability to identify its source, and by dispersed victims who generally suffer only small harms. If individual harms tend to be small, they may not be sufficient to induce citizens to initiate complaints. And if the source cannot be identified, then the polluter cannot be made to correct the problem if a complaint is filed.

Technology standards are also physically removed from the water quality problem. The physical and hydrologic linkages between field practices and water quality are difficult to ascertain at any geographic level, so the practices required in state-wide conservation plans are most often based on a best guess of what constitutes good stewardship.

In some states, technology standards are targeted to specific geographic areas, defined by a water quality problem. Monitoring plays an important role in defining the area and determining the level of action required. In many cases, the law is directed at a particular problem, such as pesticides in groundwater.

Producers in the designated areas generally must adopt specific BMPs. Enforcement is typically through inspection, making targeted technology standards more stringent than non-targeted technology standards. By focusing on specific problems in specific areas, better information on what constitutes acceptable management practices can be reasonably developed.

A good example comes from Nebraska. Nebraska is divided up into Natural Resources Districts (NRDs) which are local units of government charged with the responsibility of conservation, wise development and proper utilization of natural resources (Bishop, 1994). In 1982, the Nebraska legislature passed the Ground Water Management and Protection Act which allowed NRDs to establish groundwater control areas to address groundwater quality concerns. In 1986, the legislature gave NRDs the ability to require best management practices and education programmes to protect water quality. The best management practices defined for Nebraska were those practices that prevent or reduce present and future contamination of groundwater, and include irrigation scheduling, proper timing of fertilizer and pesticide application, and other fertilizer and pesticide management programmes.

The Central Platte NRD used this authority to develop a 'trigger' policy (Segerson, 1999) for addressing a serious and growing nitrates-in-groundwater problem. Under the Central Platte regulations, areas within the district area are divided into three phases, based on current groundwater nitrate levels. A Phase I area is defined as having an average groundwater nitrate level of between 0 and 12.5ppm (parts per million). Nitrate concentrations in Phase II areas average between 12.6 and 20ppm. A Phase III area has nitrate concentrations averaging 20.1ppm or more.

Agricultural practices are restricted according to the level of contamination. In a Phase I area, commercial fertilizer cannot be applied on sandy soils until after 1 March. Autumn and winter applications are prohibited.

Phase II regulations include the Phase I restrictions, plus the condition that commercial fertilizer is permitted on heavy soils after 1 November only if an approved nitrification inhibitor is used. In addition, all farm operators using nitrogen fertilizer must be certified by the state, irrigation water must be tested annually by farmers for nitrate concentration and the content included in fertilizer recommendations, and annual reports on nitrate applications and crop yields must be filed with the NRD.

Phase III regulations combine the Phase II requirements with requirements for split application (pre-plant and side dress) and/or nitrogen inhibitors in the spring. In addition, deep soil analysis is required annually.

An advantage of the Central Platte NRD's approach is that peer pressure can reduce enforcement costs (Randall, 1999). Having to implement ever more stringent nutrient management practices is costly to producers, so they have an incentive to monitor and 'enforce' each other to prevent 'free-riding,' thus avoiding more costly controls. Groundwater monitoring in the Central Platte NRD has shown a decrease in groundwater nitrate, indicating that the programme is working (Bishop, 1994).

Performance standards

Technology standards cover the majority of regulations. In the US, only Florida is using a performance standard to address an agricultural pollution problem. Emission-based performance standards are not generally suitable for NPS pollution, since run-off cannot be easily measured. However, in some areas such as Florida, extensive use of drainage structures allows systematic sampling to identify individual sources of agricultural pollution. The Works District Rule is being used in the area south of Lake Okeechobee to reduce the flow of phosphorus into the Everglades by placing a maximum allowable phosphorus run-off standard on dairies (Schmitz et al, 1995). Enforcement is through inspection. Dairies are allowed to reach the standard any way they can. This flexibility should result in a more efficient control of pollution than a technology standard.

Performance taxes

Performance taxes are also being applied to the Everglades in South Florida. The Everglades Forever Act calls for a uniform, per-acre tax on all cropland in the Everglades Agricultural Area. The tax starts at US$24.89 per acre per year and is increased every four years to a maximum of US$35.00 per acre unless farmers exceed an overall 25 per cent basin-wide phosphorus reduction goal (State of Florida). The tax creates the incentive to adopt BMPs and also for producers to apply pressure on recalcitrant neighbours. The number of producers is not so large that free-riding should be much of a problem.

This particular tool is flexible, in that farmers are not restricted in how they manage their operations to meet the phosphorus goal. However, the basis upon which the tax is placed, acres of cropland, is not necessarily consistent with the goal of phosphorus reduction. A more efficient approach may be to tax phosphorus loads directly.

Emission trading

A policy tool that incorporates NPSs into a regulatory programme is water quality trading. Water quality trading allows a regulated discharger to meet its Clean Water Act discharge requirements by acquiring 'credits' from other sources that take measures to reduce the regulated pollutant. Point/non-point trading takes the additional step of allowing regulated point source dischargers (factories, publicly owned treatment works) to purchase credits from unregulated non-point sources such as agriculture. One of the prerequisites of point/non-point trading markets is a regulation on point sources requiring reductions in discharges. The TMDL provision of the Clean Water Act is providing the impetus in the recent surge in interest in point/non-point trading in the US. Nutrients (nitrogen and phosphorus) are the predominant pollutants in point/non-point markets, since both point and non-point sources are major sources. The important point here is that a regulation (a cap on total emissions in a watershed) is the source of demand.

Experience with water quality trading programmes highlights the problems with non-point source-created credits. A total of 40 water quality trading programmes have been started in the US since 1990 for pollutants such as nutrients, sediment, salinity and temperature (Breetz et al, 2004). Of these, 15 include production agriculture as a potential source of credits for regulated point sources (Table 5.2). To date, trades between point and agricultural non-point sources have occurred in only four – Piasa Creek, (IL); Red Cedar River (WI); Southern Minnesota Beet Sugar; Rahr Malting (MN). Those that have occurred appear to be cost effective. For example, in the trading programme established for Rahr Malting, four NPS projects controlled phosphorus run-off at a cost of about US$2.10 per pound (based on estimated changes in phosphorus loss). Rahr Malting would have had to pay an estimated US$4–18 per pound of phosphorus reduced if it had installed pollution control equipment. However, supply-side and demand-side impediments seem to be preventing trades from occurring in most trading programmes.

Most of the impediments have to do with the characteristics of NPS pollution. Direct observation is impossible, and the performance of management practices is highly variable. This leads to a high degree of uncertainty about the number of 'credits' that a farm can provide. Point sources may be reluctant to purchase such credits. Most trading programmes try to address this uncertainty by requiring a trading ratio. A trading ratio requires more than one unit of non-

Table 5.2 *Water quality trading programmes that include agriculture*

Project	Pollutant traded	Number of trades
Cherry Creek, CO	Phosphorus	0
Lower Boise River, ID	Phosphorus	0
Piasa Creek, IL	Sediment	1
Acton, MA	Phosphorus	0
Massachusetts Estuaries Project	Nitrogen	0
Kalamazoo River, MI	Phosphorus	0
Rahr Malting, MN	Phosphorus	4
Southern Minnesota Beet Sugar, MN	Phosphorus	400
Tar-Pamlico, NC	Nitrogen, phosphorus	0
Clermont County, OH	Nitrogen, phosphorus	0
Great Miami River, OH	Nitrogen, phosphorus	0
Conestoga River, PA	Nitrogen, phosphorus	0
Fox-Wolf Basin, WI	Phosphorus	0
Red Cedar River, WI	Phosphorus	22
Chesapeake Bay Watershed	Nitrogen, phosphorus	0

Source: Breetz et al (2004)

point source discharge reduction to offset one unit point source discharge. Uncertainty ratios in water quality trading programmes generally range from 2:1 to 5:1 (CTIC, 2006). This means that a point source would have to purchase up to five units of pollutant reduction from a non-point source in order to assure that its single unit of discharge is 'covered'. The trading ratio protects the environment by ensuring adequate measures are taken, but also increases the cost of a credit, thus reducing demand.

A word on education

Education is a broad category of instruments aimed at developing an information base and improving conservation practices and programme delivery. Research and data development provide information on the economic, productivity and environmental performance of production and conservation practices. Extension and technical assistance transfer this information to farmers through education materials, demonstration projects and face-to-face contact. In the USDA, these activities are undertaken by: the Agricultural Research Service (ARS); Cooperative State Research, Education, and Extension Service; the Economic Research Service; the Agricultural Marketing Service; the Forest Service; and the Natural Resources Conservation Service.

Education by itself cannot be considered a strong tool for protecting environmental quality through conservation. The principle reason is that most of the environmental benefits occur off the farm. Education is more effective for improving productivity on the farm because the farmer can realize an economic gain. Education can, however, be an effective tool for improving environmental quality under certain conditions:

- the actions that improve environmental quality also increase profitability;
- producers have strong altruistic or stewardship motives; and/or
- the on-farm costs of environmental impairments are shown to be sufficiently large.

For example, conservation tillage increases net returns for some producers while reducing soil erosion and improving water quality. Other practices that can increase profitability and environmental quality include nutrient management and irrigation water management. Practices that improve environmental quality without boosting profits, such as filter strips and enhanced wildlife habitat, would be less likely to be adopted voluntarily without financial assistance.

Education's greatest value is as a component of an environmental improvement policy that relies on other tools such as financial incentives and direct regulation. One of the lessons learned from the USDA's Area Studies Project is that education influences which conservation practices a farmer adopts in order to meet the requirements of programme provisions such as conservation compliance (Caswell et al, 2001). By providing the information producers need to implement existing and new practices efficiently and also information about a producer's pollution contributions, overall pollution control is attained at lower cost.

Federal vs local

An important issue in developing NPS pollution control strategies is the level of government at which incentives are developed and implemented. Federal water quality law has passed responsibility for NPS control to the states. Passing responsibility for NPS pollution control to states has both advantages and disadvantages. A basic principle of the economic theory of federalism is that economic efficiency in the provision of public goods is generally best served by delegating responsibility for the provision of the good to the lowest level of government that encompasses most of the associated benefits and costs (Shortle, 1995). The impacts of NPS pollution often are most pronounced close to its point of origin. Contaminated groundwater does not move far from pollution sources. Lakes and small reservoirs are generally affected by local land uses. Likewise, streams and small rivers are impacted by land uses within relatively small watersheds. The impacts of agricultural run-off on water quality are generally most pronounced in small lakes and reservoirs, and small rivers (Goolsby and Battaglin, 1993). Also, control of NPS pollution requires regulation of land use, which traditionally has been the prerogative of states and local governments (Malik et al, 1992).

The characteristics of NPS pollution vary over geographic space, due to the great variety of farming practices, land forms, climate and hydrologic characteristics found across even relatively small areas. An efficient centralized control policy would have to account for many different situations, resulting in exceedingly high administration costs. Although reducing these costs through national standards comes at a price of reduced efficiency. An efficient decentralized policy would need to account for less variation.

Decentralized control does not easily address the problem of interstate transport of pollutants, as noted above. Cooperation would increase the likelihood of a more efficient response to pollution problems, in that a greater share of those who benefit is accounted for.

Turning responsibility for control over to the states can (and has) resulted in very uneven responses to pollution across states. States react differently to similar pollution problems for a variety of reasons. These include differences in socio-economic characteristics of a state's populace, internal partisanship, organizational capacity and the perceived severity of problems. While states may be better positioned than the federal government to develop more efficient pollution control policies, they do not always have the means or the will.

Conclusions and policy implications

The use of enforceable mechanisms to control NPS pollution has been a policy of last resort in the US. Their use is more widespread at the state level than the federal. Most states have a number of enforceable authorities that can be used to address NPS discharges. However, the coverage across types of non-point sources is often incomplete. There are also wide variations between states in the scope and nature of enforceable mechanisms. In general, the more restrictive programmes have been directed at potentially serious problems that are of immediate concern and where voluntary approaches have failed.

What is evident from the approaches that have been taken is that the characteristics of NPS pollution and the lack of specific water-quality goals have led to technology-based policies that are inherently inflexible (Table 5.3). The transactions costs of acquiring the information necessary to implement more flexible, performance-based approaches are currently too high, for both producers and regulating agencies. In the few instances where performance-based approaches are being used, unique situations have greatly reduced monitoring costs. However, performance-based approaches still face the problem of linking each individual producer's management actions to water quality outcomes.

No general statement can be made about which policy instruments give the most efficient or cost-effective control. The characteristics of NPS pollution problems vary tremendously across the country. The choice of policies to control NPS problems depends on the nature of the water quality problem, the information available to the administering agency on the linkages between farming practices and water quality, farm economics, and societal decisions about who should bear the costs of control. While technology standards are considered to be inferior to performance-based practices in this regard, this approach can give satisfactory results if the regulatory agency does a good job of identifying the right set of practices to require, and if the policy is adequately enforced. Performance-based approaches are too costly given our limited ability to link management practices to water quality.

Compliance mechanisms appear to have some favourable characteristics, and they have been effective in reducing soil erosion on highly erodible cropland in the US. However, the incentive is dependent on the level of support provided by agricultural programmes and is therefore subject to political and economic conditions. Coverage may also be incomplete. About 83 per cent of highly erodible cropland is located on farms that receive agricultural programme payments (Claassen et al, 2004). For other problems such as those related to animal waste, a much smaller percentage of producers receive payments. Compliance would not be an effective approach for addressing animal waste concerns.

A policy framework that uses a variety of tools and is based on watersheds could provide the best opportunities for cost-effective control. Research on the linkages between management practices and water quality would enable more

Table 5.3 *Characteristics of policy instruments used in the US*

Policy instrument	Effectiveness	Information costs for producers	Administration and enforcement costs	Flexibility	Where used
Compliance	Limited to programme participants, and by the level of programme support	Low	Moderate	High	Soil conservation on highly erodible land and wetland preservation
Technology standards	Good, if appropriate practices required and enforced	Low	High	Low	Some states require certain management practices in particular areas
Performance standards	Good, if farmers can link their actions to water quality performance	Medium – access to information on links to water quality	High	High	Florida only. Unique situation where run-off can be monitored relatively easily
Performance taxes	Good, if farmers can link their actions to water quality performance	Medium – access to information on links to water quality	High	High	Florida only. Unique situation where run-off can be monitored relatively easily
Emissions trading	Poor, if NPSs not included under cap	High	High	High	A number of watersheds have developed point–non-point trading programmes
Education	Poor	Low	High	High	Used extensively by USDA, can reduce costs of other instruments

widespread use of performance-based practices and the greatest opportunities for cost-effective control.

Notes

1   Two different NPS programmes arose because the Clean Water Act and Coastal Zone Management Act are under the jurisdictions of different committees in Congress.

Reference

Bishop, R. (1994) 'A local agency's approach to solving the difficult problem of nitrate in the groundwater', *Journal of Soil and Water Conservation,* vol 49, no 2, pp82–84

Braden, J. B. and Segerson, K. (1993) 'Information problems in the design of non-point-source pollution policy', in C. S. Russell and J. F. Shogren (eds) *Theory, Modeling, and*

Experience in the Management of Non-point-Source Pollution, Kluwer Academic Publishers, Boston

Breetz, H. L., Fisher-Vander, K., Garzon, L., Jacops, H., Kroetz, K. and Terry, R. (2004) 'Water Quality Trading and Offset Initiatives in the US: A Comprehensive Survey', Dartmouth College, Hanover, NH, www.dartmouth.edu/~kfv/waterqualitytradingdatabase.pdf, accessed 12 December 2006

Caswell, M., Fuglie, K., Ingram, C., Jans, S. and Kascak, C. (2001) *Adoption of Agricultural Production Practices: Lessons Learned from the U.S. Department of Agriculture Area Studies Project,* AER-792, US Department of Agriculture, Economic Research Service, Washington, DC

Claassen, R., Breneman, V., Bucholtz, S., Cattaneo, A., Johansson, R. and Morehart, M. (2004) *Environmental Compliance in U.S. Agricultural Policy,* AER-832, US Department of Agriculture, Economic Research Service, Washington, DC

Cook, K., Hug, A., Hoffman, W., Taddese, A., Hinkle, M. and Williams, C. (1991) Center for Resource Economics and National Audubon Society, statement before the Subcommittee on Environmental Protection, Committee on Environment and Public Works, US Senate, 17 July 1991

Copeland, C. and Zinn, J. A. (1986) 'Agricultural Non-point Pollution Policy: A Federal Perspective', Paper prepared for a colloquium on Agrichemical Management to Protect Water Quality, Washington, DC

CTIC (2006) *Getting Paid for Stewardship: An Agricultural Community Water Trading Guide* Conservation Technology Information Center, West Lafayette, IN

ELI (2000) *Putting the Pieces Together: State Non-point Source Enforceable Mechanisms in Context,* Environmental Law Institute, Washington, DC

Goolsby, D. A. and Battaglin, W. A. (1993) 'Occurrence, distribution, and transport of agricultural chemicals in surface waters of the Midwestern United States', in D. A. Goolsby, L. L. Boyer and G. E. Mallard (eds) *Selected Papers on Agricultural Chemicals in Water Resources of the Midcontinental United States,* Open-File Report 93-418, US Geological Survey, Reston, VA

Harrington, W., Krupnick, A. and Peskin, H. M. (1985) 'Policies for non-point source water pollution control', *Journal of Soil and Water Conservation,* vol 40, no 1, pp27–32

Lubowski, R. N., Vesterby, M., Bucholtz, S., Baez, A. and Roberts, M. J. (2006) *Major Uses of Land in the United States, 2002,* EIB-14, US Department of Agriculture, Economic Research Service, Washington, DC

Malik, A. S., Larson, B. A. and Ribaudo, M. (1992) *Agricultural Non-point Source Pollution and Economic Incentive Policies: Issues in the Reauthorization of the Clean Water Act,* AGES 9229, US Department of Agriculture, Economic Research Service, Washington, DC

Randall, A. (1999) 'Providing for the common good in an era of resurgent individualism', in F. Casey, A. Schmitz, S. Swinton and D. Zilberman (eds) *Flexible Incentives for the Adoption of Environmental Technologies in Agriculture,* Kluwer Academic Publishers, Norwell, MA

Ribaudo, M. (1989) *Water Quality Benefits from the Conservation Reserve Program,* AER-606, US Department of Agriculture, Economic Research Service, Washington, DC

Ribaudo, M. O., Horan, R. D. and Smith, M. E. (1999) *Economics of Water Quality Protection from Non-point Sources: Theory and Practice,* AER-782, US Department of Agriculture, Economic Research Service, Washington, DC

Ribaudo, M., Gollehon, N., Aillery, M., Kaplan, J., Johansson, R., Agapoff, J., Christensen, L., Breneman, V. and Peters, M. (2003) *Manure Management for Water Quality: Costs to Animal Feeding Operations of Applying Manure Nutrients to Land,* AER-824, US Department of Agriculture, Economic Research Service, Washington, DC

Schmitz, A., Boggess, W. G. and Tefertiller, K. (1995) 'Regulations: Evidence from the Florida dairy industry', *American Journal of Agricultural Economics,* vol 77, no 5, pp1166–1171

Segerson, K. (1999) 'Flexible incentives: A unifying framework for policy analysis,' in F. Casey, A. Schmitz, S. Swinton and D. Zilberman (eds) *Flexible Incentives for the Adoption of Environmental Technologies in Agriculture,* Kluwer Academic Publishers, Norwell, MA

Shortle, J. (1995) 'Environmental federalism: The case of US agriculture?', in J. R. Braden, H. Folmer and T. Ulen (eds) *Environmental Policy with Economic and Political Integration: The European Union and the United States,* Edward Elgar, Cheltenham

Smith, R. A., Schwarz, G. E. and Alexander, R. B. (1994) *Regional Estimates of the Amount of US Agricultural Land Located in Watersheds with Poor Water Quality,* Open-File Report 94-399, US Department of the Interior, US Geological Survey, Reston, VA

Smith, R. A., Alexander, R. B., Schwarz, G. E. and Ieradi, M. C. (2005) 'Effects of Structural Changes in US Animal Agriculture on Fecal Coliform Contamination of Streams', Poster presentation, US Department of Interior, US Geological Survey, Reston, VA

USEPA (1988) *Non-point Sources: Agenda for the Future,* US Environmental Protection Agency, Office of Water, Washington, DC

USEPA (2002) *National Water Quality Inventory: 2000 Report to Congress,* EPA841-R-02-001, US Environmental Protection Agency, Office of Water, Washington, DC

USEPA (2005) *National Section 303(d) List Fact Sheet* http://oaspub.epa.gov/waters/national_rept.control#IMP_STATE, US Environmental Protection Agency, accessed 3 January 2008

Wicker, W. (1979) 'Enforcement of Section 208 of the Federal Water Pollution Control Act Amendments of 1972 to control non-point source pollution', *Land and Water Law Review,* vol 14, no 2, pp419–446

6

Non-point Pollution Control: Experience and Observations from Australia

Michael D. Young

Introduction

This chapter focuses on the techniques used to manage non-point sources of pollution associated with irrigation in Australia. In the interests of brevity, we ignore the many opportunities to reduce non-point sources of pollution by, for example, charging for the full cost of providing water, unbundling water rights and allowing irrigators to trade water entitlements and water allocations.

Arguably, and in terms of economic impact, there are two main non-point pollution problems in Australia. The first of these is irrigation salinity in the Southern Connected River Murray system. The second is sediment and nutrient pollution on the Great Barrier Reef in Queensland. Another lesser problem, usually found in association with irrigated pastures for dairying, is nitrate contamination of groundwater and surface water systems. Pollution of waterways by pesticides is now considered to be a problem that is well under control (Radcliffe, 2002).

Experts on approaches to the regulation and control of non-point sources of pollution in Australia often begin by drawing attention to the way salinity is managed by the Murray–Darling Basin Commission and, more recently, to the approach that the town of Busselton in Western Australia has taken to the control of nitrate and phosphate pollution in the Geographe Bay.

While it is a point rather than a non-point pollution control scheme, many experts would also draw attention to the highly innovative and well-developed salinity trading programme that has been operating for many years in the Hunter River in New South Wales.

The last initiative that Australian non-point source pollution managers tend to draw attention to is the last two decades of institution reform that have tended to devolve responsibility to local and regional organizations and to make greater use of market-based mechanisms.

As background one may also observe that, as a result of the recent drought, the focus of irrigation policy development in Australia has shifted from water quality issues to water quantity and allocation issues. As a general rule, non-point source pollution in Australia appears to be inversely related to fluctuations in seasonal rainfall. At the time of writing, in May 2008, almost all policy attention in Australia is focused on water quantity issues. Behind the scenes, however, there is awareness that when the drought breaks non-point sources of pollution problems could return with a vengeance.

Institutional approaches

Australia is a federation of states and its constitution leaves most responsibility for the management of non-point pollution issues with individual state governments. For many issues, like the registration and control of pesticide use, the degree of state cooperation is high. A national pesticide registration scheme is in place and states are responsible for controlling use within their area (Radcliffe, 2002).

In all states, it is usual for regulations to set a minimum standard and impose an environmental duty of care on land users. In South Australia, for example, environmental protection legislation requires land users to take all reasonable steps to prevent or minimize environmental harm (Box 6.1).

The main advantage of this duty of care approach is that it shifts the onus of responsibility from one that requires a person to comply with legislation, to one that requires the pesticide user to take all reasonable steps to avoid causing harm (Young et al, 2003). When new information emerges and or an industry organization sets a standard – even a voluntary one – all users are expected to comply with this new standard.

Education and governance

From an irrigation perspective, the main non-point pollution control technique is to control or regulate land use and where possible begin the process of shifting community attitudes by providing incentives to early adopters. These programmes are now managed primarily by a suite of Natural Resource Management Boards or 'catchment boards' as they are often called. Driven by a National Action Plan for Salinity and Water Quality (CoAG, 2000), 56 Natural

---

BOX 6.1 EXTRACT FROM THE ENVIRONMENTAL PROTECTION ACT ESTABLISHING A GENERAL ENVIRONMENTAL DUTY OF CARE ON ALL LAND USERS

25 – General environmental duty

(1) A person must not undertake an activity that pollutes, or might pollute, the environment unless the person takes all reasonable and practicable measures to prevent or minimize any resulting environmental harm.

(2) In determining what measures are required to be taken under subsection (1), regard is to be had, amongst other things, to –
    (a) the nature of the pollution or potential pollution and the sensitivity of the receiving environment; and
    (b) the financial implications of the various measures that might be taken as those implications relate to the class of persons undertaking activities of the same or a similar kind; and
    (c) the current state of technical knowledge and likelihood of successful application of the various measures that might be taken.

(3) In any proceedings (civil or criminal), where it is alleged that a person failed to comply with the duty under this section by polluting the environment, it will be a defence –
    (a) if –
        (i) maximum pollution levels were fixed for the particular pollutant and form of pollution concerned by mandatory provisions of an environment protection policy or conditions of an environmental authorization held by the person, or both; and
        (ii) it is proved that the person did not by so polluting the environment contravene the mandatory provisions or conditions; or
    (b) if –
        (i) an environment protection policy or conditions of an environmental authorization provided that compliance with specified provisions of the policy or with specified conditions of the authorization would satisfy the duty under this section in relation to the form of pollution concerned; and
        (ii) it is proved that the person complied with the provisions or with such conditions of an environmental authorization held by the person.

(4) Failure to comply with the duty under this section does not of itself constitute an offence, but –
    (a) compliance with the duty may be enforced by the issuing of an environment protection order; and
    (b) a clean-up order or clean-up authorization may be issued, or an order may be made by the Environment, Resources and Development Court under Part 11, in respect of non-compliance with the duty; and
    (c) failure to comply with the duty will be taken to be a contravention of this Act for the purposes of section 135.

Source: Environmental Protection Act (1993) Government of South Australia

---

Resource Management Regions have been established across Australia in a manner that has produced complete coverage of the country (see Figure 6.1).

In each Australian Natural Resource Management Region, a board has been established with responsibility, in partnership with the state and the federal

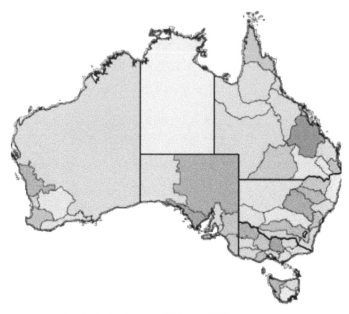

Source: www.nrm.gov.au/nrm/region.html, accessed 25 January 2008

Figure 6.1 *Distribution of Australia's 56 Natural Resource Management Regions*

government, for managing and protecting the natural resources of the region. Boards are made up of a mixture of local resource users, community and government representatives. Each board is able to employ its own staff and, in a number of cases, staff have been seconded from state departments to these boards.

In each case, one of the board's first responsibilities is to establish, and have accredited, a Natural Resource Management Plan. Once the plan is accredited, the board gains access to money made available to it via the national government's Natural Heritage Trust and its National Action Plan for Salinity and Water Quality.[1]

By way of example, the South Australian Murray–Darling Basin Natural Resource Management Board has been assigned responsibility for the development of the water allocation plan for 'its' part of the river. In addition to this and many other functions, the board tries to act proactively. One of the ways that boards, like this one, seek to take a proactive role is to employ staff to assist irrigators to prepare irrigation development management plans. These plans must be prepared before permission can be obtained to irrigate a new area of land and must show how non-point sources of pollution will be managed. Among other things, the irrigation development plans are required to be prepared in a manner that is consistent with Local Water Management Plans developed by boards in consultation with the local landholders.

Market-based approaches to regulation and delivery of non-point pollution outcomes

One of the more innovative features of the National Action Plan for Salinity and Water Quality was a cooperative decision by all states and the Commonwealth government to fund a Aus$10 million trial of the use of market-based instruments as a means to improve resource use.[2] Differing from more conventional approaches that tend to regulate the use of input and land use practice, market-based instrument approaches enable resource users to be much more innovative and outcome focused.

The first round of these trials involved expenditure of Aus$5 million on 11 trials and has now been completed. The second round involves a further nine trials that will spend a further Aus$5 million.

The result of these trials and the earlier experience has been the emergence of strong national interest in using market-based instruments (MBIs) as part of the mix of instruments used to manage non-point sources of pollution. In his independent review of the first round of this programme, Grafton (2005) found that:

- auctions, cap and trade (for point sources) and offsets can be successfully used to address a wide variety of water quality, salinity and environmental problems in the Australian landscape;
- MBIs, especially auctions, can deliver large cost savings relative to traditional natural resource management;
- to effectively implement MBIs there needs to be very good biophysical modelling at the farm or paddock level, and adequate monitoring and enforcement of landholders' actions;
- to generate cost savings, MBIs require adequate testing and adaptation prior to implementation and well-developed communication strategies to maximize participation by landholders; and
- there is no one-size-fits-all approach to environmental problems, and MBIs will need to be tailored and adapted to particular circumstances.

Ongoing trials are now focusing on the development of offset trading schemes, improvement of benefit indices used in many of these schemes, options for improving participation rates in voluntary programmes and increasing cost effectiveness per unit of outcome delivered. As indicated below, offset programmes are proving to be effective in helping to manage non-point sources of pollution in areas where the on-site impacts of existing problems are such that it is profitable for landholders to trade their water allocation or shift their water entitlement to a new area.

Salinity trading in the Murray–Darling Basin

One of the first market-based approaches to the management of non-point pollution is the salinity credit and debit programme operated by the Murray–Darling Basin Commission. Under this agreement, the Queensland, Victorian, New South Wales and South Australian governments have agreed to a set of end of valley targets and, also, to keep river salinity at Morgan on the River Murray in South Australia below 800μS/cm of electrical conductivity (EC) for 95 per cent of the time.[3]

In order to maintain river salinity below 800EC in the most efficient way possible, each state has agreed to account for and pay for its contribution to the problem. Each year, estimates are made of the likely increase in salinity as a result of new and current land use against a modelled 1975–2000 benchmark. These estimates are then recorded as credits and debits on 'A' and 'B' salinity registers. The 'A' salinity register is used to record the impact of all recent causes of change in river salinity, and the 'B' salinity register to account for impacts due to changes in land use that occurred before an agreement to manage salinity was established.

States then invest collectively in the construction of salinity interception and other schemes designed to keep river salinity within acceptable levels. Debits to the 'A' salinity register are charged to states according to an estimate of the economic impact of each unit of salinity as measured at Morgan. Credits are earned by investing in salinity interception schemes and/or ceasing irrigation in areas that contribute to the river salinity problem. In order to keep river salinity within reasonable levels, states are then required to fund the construction of salinity interception schemes across the entire basin in proportion to their balance on the salinity register. The result is a regime that enables each state to offset its 'A' salinity register costs at much less cost than would otherwise be the case if it could only offset salinity within its own state.[4]

Enhancements of this debit and credit salinity management system that are under consideration include extension of the scheme to allow individual irrigators to sell credits into the system, the introduction of the provision of dilution flows as a means to keep salinity within acceptable levels and a requirement for each scheme to hold a water entitlement. Benefit/cost assessments are performed before each new salinity interception scheme is put in place and the methodology used to make these assessments periodically reviewed.

Hunter River salinity trading

Another well-known Australian salinity trading scheme is that operated in the Hunter River in New South Wales. While, at this stage in its development, it is a point-based salinity trading scheme, it is unique in that it involves the use of a tradable credit system that enables firms to manage saline wastewater and decide when to discharge saline wastewater into the river. At the start of each year, each firm is issued with discharge permits that they can either use or sell to firms that have run out of storage capacity and need to release some saline wastewater into the river.

Provided they have enough permits, firms are free to determine when they discharge their wastewater. The number of permits required to discharge a unit of salt, however, is a function of ambient river salinity. As ambient salinity increases, more and more discharge permits are required. Thus, in times when ambient river salinity levels are high, firms have an incentive to store saline wastewater and when ambient salinity is low, they have an incentive to discharge. In the early stages of this programme, trading among the firms involved was rare but, as experience has developed, trading has become more common (Anon, 2002).

While this scheme operates only among point source emitters, it does show that market-based instrument approaches can be operated in a manner that provides incentives for people to take account of ambient water quality.

Busselton bubble licensing

On the other side of Australia, an interesting point source to non-point source trade-off has been made in association with the town of Busselton, south of Perth in Western Australia. Busselton is next to a large but closed bay known as Geographe Bay which, unfortunately, has a limited capacity to assimilate nitrates and phosphates and is surrounded by a dairy industry that was responsible for 95 per cent of the nitrates and phosphates that flow into the bay.

A growing population meant, however, that the town had to find a way to dispose of more and more treated sewage that is rich in nitrates and phosphates. In the past, all of Busselton's treated sewage water was released into Geographe Bay but when the town applied for permission to increase the amount of nitrates and phosphates it deposited into the Bay, it was advised that it would need to find another solution as Geographe Bay had no further capacity to absorb nitrates and phosphates. After appropriate analysis, it was recommended that the town establish a woodlot where wastewater could be spread and assimilated into the environment. Further analysis then revealed that it would cost Aus$4 million to establish the woodlot and then around Aus$500,000 per annum in running costs.

Those involved, however, suspected that it would be cheaper to reduce the load coming from the dairy industry and, in return, obtain permission to dispose of nearly 3 million litres of treated sewage water into Geographe Bay in partnership with the surrounding dairy industry – 95 per cent of the nutrient load into Geographe Bay came from surrounding rural catchments and only 5 per cent from the town (O'Grady and Humphries, 2003).

Box 6.2 Extract from Busselton Environmental Improvement Initiative final report

The first two years of the program were considered largely unsuccessful, generating only 13 acceptable projects to the value of $90,000 of a possible $400,000 for the two years. With guidance from the Local Advisory Group, the State Steering Committee and participating farmers, problems were identified and improvements made to the operation of the Environmental Improvement Initiative (EII). These improvements included the hiring of an environmental engineer specializing in rural wastewater management; establishing a loan agreement for farmers who could not meet their 50% contribution up front; including the Carbunup catchment to increase the pool of applicants; and abolishing individual funding rounds.

Changes to the EII model and to the dairy industry in 2002 had a significant effect on the uptake of EII funds in the remaining three years. By the end of the project there were 62 signed agreements and $832,324 of EII grant funding spent on nutrient management projects and associated research and consultancy. Over $1.2 million was spent in matching funds by project participants.

Nutrient management projects included: 30 dairy waste management projects, 1 dairy composting project, 6 potato growers' fertilizer management projects, 1 groundwater denitrification trench, 1 community landcare nursery, 21km of waterway fencing, a nutrient management project, 25ha of perennial pastures, 1 beef feedlot waste management project, 2 catchment friendly workshops and 5 consultancy and research projects.

A total of 70 applications for funding were submitted over the 5 years. Of these, 57 were approved, with 13 applicants cancelling their funding and 44 completing their projects. Applications were cancelled due to a variety of reasons including ceasing operation or not having the funds to support the project. The approved applications include several applicants who applied for a second or third grant for staged works to dairy waste management projects.

The nutrient loss and waste reduction projects funded by the EII over the 5 years achieved a total estimated load reduction of 73.5 tonnes of nitrogen and 18 tonnes of phosphorus each year.

The EII was considered a more cost effective outcome to reduce the nutrient load into Geographe Bay than the woodlot. The significant cost of $4 million for the woodlot would need to manage 29 tonnes of total nitrogen and 4 tonnes of total phosphorus per year. $1 million spent on EII projects managed 73.5 tonnes of total nitrogen and 18 tonnes of total phosphorus. A simple comparison estimates the EII to be 12 times more cost effective for nitrogen and 21 times more cost effective for phosphorus.

The initial lack of uptake of EII funding highlighted that, in the absence of any clear financial and business benefits, regulatory and enforcement mechanisms, supported by technical and financial assistance, are required to drive change in environmental management practices.

Source: McGuire et al (2007)

After considering the options, and in recognition of the economic and environmental consequences of taking this alternative approach, the Western Australian Environmental Protection Authority approved what is, in effect, a bubble licence that authorized an increase in emissions from Busselton on the condition that it implement an Environmental Improvement Initiative that would spend Aus$1 million assisting landholders to reduce contamination of surface and groundwater systems that feed into Geographe Bay (McGuire et al, 2007).

In the first two years, the initiative relied on voluntary submissions for funding and was a failure, so in the third year a specialist environmental consultant was employed to make direct one-to-one contact with farmers. Under the revised initiative, farmers are offered low interest rate loans to improve their management systems, with half of the interest rate cost of these loans met by the sewage treatment plant and half by participants (O'Grady and Humphries, 2003; McGuire et al, 2007). As indicated in Box 6.2, the final result proved to be 12 times more cost effective for nitrogen management and 21 times more cost effective for phosphorus management.

Dryland salinity control

The next example of Australian experience in the use of market-based instruments to control non-point sources of pollution comes from the catchment of Bet Bet in the state of Victoria. Bet Bet is a mixed farming area that is one of the largest contributors to salinity in the Southern Connected River Murray system. Given this, it was decided under the market-based instrument trial programme to assess whether or not a standard cap and trade model could be used to reduce salinity impact on a river system from dryland agriculture.

As it was a trial, the approach taken was to invite farmers to tender to participate in a programme that would enable them to trade salinity reduction credits in order to deliver agreed programme outcomes more cost effectively. As explained by Connor et al (2008b):

> … *in lieu of extant specified property rights, participants were invited to agree to obligations to provide groundwater recharge credits in exchange for pecuniary compensation. Participants were able to meet their obligations to supply groundwater recharge credits through land management actions resulting in monitored outcomes consistent with contractual obligations to reduce recharge. Alternatively, those in deficit were provided the option to obtain sufficient credits through market exchange. Surplus transferable recharge credits were produced by those participants who exceeded their own contractual obligations through improved land management.*

Although the project is still to be completed, early assessments have shown it to be an extremely cost-effective way of reducing dryland salinity impacts.

In essence, the main approach taken was to enter into contracts to reduce the expected impacts of land use practice with landholders and then allow them to deliver the contracted outcome in the most efficient way possible. The feature of this project that differentiates it from every other project of its kind is the use of collective performance incentive payments. Under this arrangement, in addition to individual payments, a reward payment is made to all participants if and only if the trial delivers the agreed outcome.

Although still under way, several observations can be made about this trial. The first is that the specification of required outputs from a region rather than the direct control of land use practice is much more efficient and cost effective. The second is that the use of a collective group incentive payment (a reward for making the programme achieve its goal) increases community interest and participation. A third important observation is that collective group incentive payment arrangements give all participants an incentive to fix any design loopholes that emerge rather than to seek ways to profit by exploiting them (Connor et al, 2008b).

Environmental benefit indices

In recent years, Australia has also begun to use tenders as a means to achieve more conservation and reduce river pollution at less cost. In every case, the approach taken to develop an environmental benefit index and then invite landholders to bid to supply environmental benefits on the understanding that they will only be selected to participate in such a scheme if their bid is cost competitive per unit of environmental benefit offered. Early experience with the use and development of these schemes suggested dramatic returns per public dollar invested and this has led to an expansion of effort in this area.

One of the most recent trials has been implemented in the Onkaparinga Catchment in South Australia where landholders have been contracted to both improve the health of watercourses and riparian systems. Differing from earlier tenders, such as those developed by Stoneham et al (2003), this tender was able to collect information on the relative merits of different tender methodologies. The main finding from the Onkaparinga water quality tender is that most of the benefits derive from the use of a benefit index in programme selection, rather than from the use of a tender process to select participants. A second, counter-intuitive finding is that such programmes are likely to be much more cost effective if all participants bid in the knowledge that they will receive a uniform payment per unit of benefit delivered rather than each being paid only the amount per unit they bid. The reason for this is that under the uniform bid approach, the optimal strategy is to reveal the minimum bid you would accept as there is no opportunity to profit by nominating a higher bid price (Connor et al, 2008b).

Pricing, charging and trading

In parallel with the above policy approaches, the Victorian government is using the combination of water trading and charging approaches to reduce river salinity. While the introduction of water trading has caused many other problems, from a non-point source of pollution perspective, the general experience has been that it has encouraged the movement of irrigation from high to low impact areas. In the Kerang–Pyramid Hill–Boort region, for example, community

Table 6.1 *Summary of salinity levy payable for permanent trades (Perm) and temporary trades (Temp) in Victoria from Nyah to the border for Low Impact Zones (LIZ). No trade is allowed within or into a High Impact Zone (HIZ)*

Trade from	Trade to									
	LIZ 1		LIZ 2		LIZ 3		LIZ 4		HIZ	
	Temp	Perm	Temp	Perm	Temp	Perm	Temp	Perm	Temp	Perm
Outside area	$2.60	$26.00	$6.50	$65.00	$13.00	$130.00	$26.00	$260.00	No trade	No trade
LIZ 1	$0.00	$0.00	$3.90	$39.00	$10.40	$104.00	$32.40	$234.00	No trade	No trade
LIZ 2	$0.00	$0.00	$0.00	$0.00	$6.50	$65.00	$19.50	$195.00	No trade	No trade
LIZ 3	$0.00	$0.00	$0.00	$0.00	$0.00	$0.00	$13.00	$130.00	No trade	No trade
LIZ 4	$0.00	$0.00	$0.00	$0.00	$0.00	$0.00	$0.00	$0.00	No trade	No trade
HIZ	No trade	No trade	No trade	No trade	No trade	No trade	No trade	No trade	No trade	No trade

Source: Lower Murray Urban and Rural Water, www.srwa.org.au/index2.htm

efforts to reduce salinity impacts using collective planning approaches reduced river salinity at Morgan by approximately 6EC. When water trading was introduced, however, the benefits of being able to move salinity to more productive locations produced a further 20EC of salinity at Morgan (Young et al, 2006).

Pushing the trading policy opportunity further, Victoria had zoned areas where irrigation salinity is a problem into a series of high and low impact areas. Trading water into a high impact area is prohibited and, as summarized in Table 6.1, a salinity levy is used to help fund Victoria's contribution to the salinity debit and credit system discussed earlier in this chapter. The result is a powerful signal that makes all involved in and or considering trading water to take greater account of the impact of the trades they make on river salinity.

Salinity offset trading in South Australia

In contrast with Victoria's pricing approach, South Australia is using an offset approach to manage the salinity impacts of new irrigation development. In this state, control of non-point sources of pollution is achieved by making the relocation of irrigation from one area to another conditional upon a requirement that the relocation has no net impact on water quality.

In order to facilitate salinity offset trading in South Australia, all areas that could be used for irrigation have been classified into one of three zones: low impact zones; high impact zones; and high impact zones that are protected by a salinity interception scheme. As a general rule in high salinity impact areas, applications for a licence to irrigate will be approved only if the impact of the proposed development is offset by, for example, decommissioning irrigation elsewhere.[5]

The result has been a significant increase in opportunities to develop irrigation and no cost to the government or the community.

Conclusion and policy implications

Australian experience suggests that a regulatory floor will always be necessary as part of the mix of policies used to control non-point sources of pollution. It also suggests that when adopting regulations there are merits in establishing a duty of care on land users and focusing on expected outcomes rather than the excessive control of inputs and practices.

The main conclusion from this chapter is that countries should consider using market-based approaches to assist with the control of non-point sources of pollution in irrigation areas and elsewhere. The main advantage of market-based approaches is that they offer greater flexibility in achieving control, leave greater opportunity for innovation and allow non-point sources of pollution to be controlled at less cost.

Market-based approaches can operate both at a government-to-government level and at a farm level. When implemented at farm level, Australian experience suggests that the main opportunity for improvement occurs during the process when an area is being irrigated for the first time and then only if the landholder is required to obtain permission to irrigate an area of land.

When market-based approaches to non-point pollution control are implemented at the farm level, enforcement is a major problem and the costs of implementing such a scheme can be prohibitive. One option being trialled successfully in Australia is the use of devolved administrative systems that provide incentive payments (cash rewards) to participants when an aggregate pollution-reduction target is achieved.

Another finding is that considerable gains can be achieved through the development of environmental benefit indices and their use in deciding where and how to invest funds.

The last area of opportunity identified in this chapter is the opportunity to combine the management of point and non-point sources of pollution. It is the total amount of pollution that matters, not where the pollution comes from. When the cost of controlling the next unit of point-source pollution is several times more expensive than the cost of reducing a unit of non-point pollution, consider setting up a point/non-point pollution trading or offset scheme. The result can be a cost-effective win for all.

Notes

1　A new national government was elected late in 2007 and it has recently announced a new Caring for Country Program that will replace the existing array of Natural Heritage Trust and other related programmes. It is expected that the result will be a suite of arrangements that are more competitive and seek to produce greater results per dollar invested.
2　For more information on this trial go to www.napswq.gov.au/mbi/index.html.
3　Electrical conductivity measured in μS/cm can be generally converted to mg/litre of total dissolved salts using a conversion factor of 0.6 (MDBC, 1999).
4　A fuller description of the scheme is available at www.mdbc.gov.au/salinity/basin_salinity_management_strategy_20012015.
5　Development is normally approved in low salinity impact zones provided South Australia has salinity credits that are available on the 'A' register described above. Development is possible also in high salinity impact areas that are protected by a salinity interception scheme. For more information on this scheme see www.dwlbc.sa.gov.au/murray/salinity/zoning.html.

References

Anon (2002) 'Investigating New Approaches: A Review of Natural Resource Management Pilots and Programs in Australia that Use Market-based Instruments', www.napswq.gov.au/publications/books/mbi/pubs/pilot-program-review.pdf

CoAG (2000) *Intergovernmental Agreement on a National Action Plan for Salinity and Water Quality*, Council of Australian Governments, Government of Australia

Connor, J., Ward, J., and Bryan, B. (2008a) 'Exploring the cost effectiveness of land conservation auctions and payment policies', *Australian Journal of Agricultural and Resource Economics*, vol 52, no 3, pp303–319

Connor, J., Ward, J., Clifton, C., Proctor, W. and Hatton MacDonald, D. (2008b) 'Designing, testing and implementing a trial dryland salinity credit trade scheme', *Ecological Economics*, vol 67, no 4, pp574–588

Grafton, Q. (2005) *Evaluation of Round One of the Market Based Instrument Pilot Program*, Report to the National MBI Working Group, Canberra, www.napswq.gov.au/publications/books/mbi/pubs/round1-evaluation.pdf, accessed 1 February 2008

McGuire, L., Newman, L. and Humphries, R. (2007) *Busselton Environmental Improvement Initiative 2000-2004*, Water Corporation, Perth, www.watercorporation.com.au/_files/BEII_Final_Report.pdf, accessed 1 February 2008

MDBC (1999) *The Salinity Audit of the Murray Darling Basin, A 100 Year Perspective*, Murray–Darling Basin Commission, Canberra

O'Grady, B. and Humphries, R. (2003) 'Confronting the Failure of Voluntarism in River Basin Management: The Water Corporation's Busselton Environmental Improvement Initiative', Paper presented to River Symposium 2003, www.watercorporation.com.au/_files/BEII_Riversymposium2003.pdf, accessed 1 February 2008

Radcliffe, J. C. (ed) (2002) *Pesticide Use in Australia*, Australian Academy of Technological Sciences and Engineering, Melbourne, 310pp

Stoneham, G., Chaudhri, V., Ha, A. and Strappazzon, L. (2003) 'Auctions for conservation contracts: An empirical examination of Victoria's BushTender Trial', *The Australian Journal of Agricultural and Resource Economics*, vol 47, no 4, pp477–500

Young, M., Shi, T. and Crosthwaite, J. (2003) *Duty of Care: An Instrument for Increasing the Effectiveness of Catchment Management*, Department of Sustainability and the Environment, Melbourne

Young, M. D., Shi, T. and McIntyre, W. (2006) *Informing Reform: Scoping the Affects, Effects and Effectiveness of High Level Water Policy Reforms on Irrigation Investment and Practice in Four Irrigation Areas,* Collaborative Research Centre for Irrigation Futures, Adelaide Technical Report No 02/06 available at www.irrigationfutures.org.au/newsDownload.asp?ID=295&doc=CRCIF-TR-0206-col.pdf

7

China's Water Issues: Transition, Governance and Innovation

Yi Wang

Introduction

Water is the foundation on which human civilization depends for survival and development, and water management is, therefore, an eternal subject of study in human society. In a traditional society, water management concerned the survival of the nation and the destiny of the state; in a modern society, water is the core element of sustainability of human civilization. In Chinese history, governing the country has always been closely associated with the governance of water resources, just as Guan Zi[1] says: 'Floods and droughts control are fundamental to govern a country' (Gu, 2006, p5). Because of the importance of water management, traditional Chinese society is sometimes called a 'hydraulic society' (Wittfogel, 1957).[2]

Water governance has had a far-reaching impact on shaping China's traditional social and political structure. China has accumulated a wealth of experience in water management over thousands of years and there are many successful experiences and model hydraulic engineering projects. From 'Da Yu fighting against floods', which marked the start of the ancient state, and the 'Du Jiang Yan Water Works' which has been irrigating the Chengdu Plain for 2000 years, to the relatively complete water management system at central government level, all have demonstrated the wisdom and experience of the Chinese nation in controlling water. In traditional Chinese society, water management focused on flood prevention, water channel management and irrigation. Such focuses became the fundamental responsibility of the central government, and each government department and local authority was assigned corresponding functions.

The Yellow River basin gave birth to and brought up the Chinese civilization; however, modern rivers suffer from human activities. The Chinese civilization has lasted for thousands of years in that, on the one hand, it has benefited from the vast and easy-to-till land in the Yellow River basin (Ge, 2005) and, on the other hand, it is associated with our respect for and ability to use natural laws. The Zhengguo Canal, dug more than 2000 years ago, still benefits the region thanks to the construction efforts and decades of appropriate maintenance. With social development, people have gradually learned to build dykes and prepare channels to control water and sand. All these methods have, to a certain extent, helped to ease the threat of floods of the Yellow River. However, the Chinese have overstressed the role of man and ignored the laws governing the river since 1950. Following the population expansion and fast economic growth of modern China, rivers have already been overtapped. Depleted river sources, reduced water flows, polluted water bodies and deteriorating freshwater ecosystems are threatening the development of river basins and the continuation of riparian civilization.

With the acceleration of industrialization since the 1980s, China has encountered new challenges in water issues. Compared with the traditional agrarian society, the water problems confronting China now are undergoing major changes. First, China faces multifaceted crises and challenges in water shortage, water pollution, freshwater ecosystem degradation, and floods and droughts, which are more complicated than the traditional problems of floods and irrigation. Second, all the problems have been extended from local or parts of river sections to entire basins or regions, and could even lead to a global impact. And third, each problem has different types of transition making it more complicated (Wang, 2002, 2006).

China's per capita water availability is diminishing. The cause of water shortage is no longer a simple issue of quantity, but a combination of inadequate water supply, low efficiency and pollution. Water demand in the future will continue to grow and this trend is projected to continue until 2030 (Sustainable Development Strategy Study Group of the Chinese Academy of Sciences, 2007) (Table 7.1).

Table 7.1 *Trends in water use in China (billion m³)*

Year	Total	Agricultural	Industrial	Municipal
1980	440.8	371.6	41.8	27.4
1990	486.8	376.4	69.2	41.2
2000	563.0	386.0	116.0	61.0
2005	563.3	357.8	128.6	76.9
2030	653.5	392.1	156.8	104.6

Source: Sustainable Development Strategy Study Group of the Chinese Academy of Sciences (2007, pp20, 29)

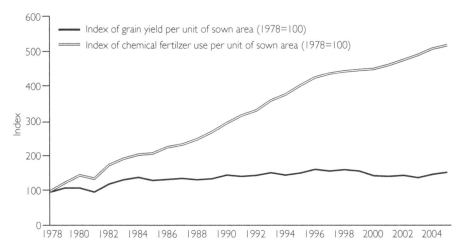

Figure 7.1 *Trends of grain yield and chemical fertilizer use per unit of sown area (1978–2005)*

Total water pollutant discharge is increasing, but the type of water pollution is also changing from traditional pollution with conventional pollutants to a compound type of pollution with new and old pollutants interacting with each other and also from industrial pollution-dominant to sewage pollution-dominant. In addition, there is the coexistence of industrial pollution and non-point source pollution from agriculture. Eutrophication of water bodies is becoming increasingly serious. In 2005, of the seven major waterway systems in China, the water quality in 27 per cent of the 411 surface water monitoring cross-sections failed to reach the standard of Category V^3 – which means they are becoming unusable. All the river sections running through cities are polluted. More than 80 per cent of the lakes in the east and south-west of the country have become eutrophicated to varying degrees mainly due to non-point source pollution. The increasing use of fertilizers will create more and more stress from non-point source pollution (Figure 7.1). About 300 million people do not have access to safe drinking water. The groundwater in some areas is seriously polluted. In general, the deterioration of water quality is not under control. The pollution load has gone far beyond the water environmental capacity (Sheng, 2006a). The prospects for pollution prevention and control are extremely pessimistic.

The excessive development of water resources, worsening water pollution and poor management of irrigation facilities have all contributed to low river flow, shrinking lakes, diminishing wetlands, sinking ground, seawater intrusion and reduction in aquatic species. The freshwater ecosystem functions will continue to degrade even though improvements occur in some places.

With regard to floods and droughts, the standards for flood prevention in major rivers are low (generally resisting floods for 20–50 years); losses from droughts and floods have increased year by year (Sustainable Development

Strategy Study Group of the Chinese Academy of Sciences, 2007, p6); and the risks of extreme weather events have increased and the threat of such events will exist for a long time.

Global warming has led to an increased frequency of droughts and floods in northern and southern China, respectively (The Editorial Board of China's National Assessment Report on Climate Change, 2007). The trend of climate change will exacerbate the water shortages in northern China. If the speed of global warming is further accelerated, it may exert a more significant negative impact on agricultural and livestock production and water supply.

The water crises will become complicated and multifaceted in the long run as a result of the interaction and overlapping of the above issues. The growing population, causing fast and intensive social and economic development, will have to rely on scarce water resources, limited water environmental capacity and fragile freshwater ecosystems.

All the evidence suggests that, of the above problems, water pollution has become a key issue that will have a major bearing on the present stage of social and economic development and on other water problems, and is awaiting urgent solution. If water pollution is not controlled effectively, all the efforts to raise water supply capacity, protect the water resources and mitigate the impact of floods and droughts – no matter how great the achievements – are likely to be nullified by the worsening water pollution and the consequential huge economic losses.

Concomitant with the changes of water problems is the major transition of water governance models. The current water governance structure and management system has evolved over the years and, although it has made significant achievements over the past half century, it is incapable of meeting the current water challenges. The key challenges include fragmented water quality and quantity management, competition between economic development and environmental protection, uncoordinated cross-sectoral and trans-jurisdiction development and the conflicts of different interest groups. So the water crises the Chinese face are, superficially, resources and environmental crises; but in essence, it is a governance crisis. We must change the traditional water management system[4] and establish a modern water governance and integrated water resources management system in response to the current situation.[5]

Transition in the water governance model is also an objective demand of general social and economic transition. Since the 1980s, China has been marching from a planned economy towards a market economy. The acceleration of industrialization and urbanization has enabled China to shake off the bondage of agrarian society at state level. The interest groups have become more and more diversified and influential; constitutional democratic polity is being established. But our ideas about governments are still in the agrarian society. The tradition of unitary, highly centralized policy decision making and ideas prevailing in the planned economy are still affecting the solution of the water problems. In the face of new challenges and against the general backdrop of economic transition, China is undergoing changes with regard to ideas of water governance, changing towards the general direction of respecting the natural law, pursuing harmony between man and water, improving governance and realizing innovation in all areas (Wang, 2003).

Serious water issues, water governance crisis and difficulties in achieving the targets

Management failure is a root cause of the multiple problems, and government functions are in urgent need of adjustment. China's water problems are the result of the dual action of natural factors and human activities. Over the past decades, China tended to stress technology and engineering solutions and neglect institutional and management measures. However, many water problems, including the cut-off of the Yellow River flow, man-made segmentation of river hydrological regimes, incomplete data and information about pollution sources and discharge, and improper responses to water-related emergencies, are closely associated with inappropriate institutional arrangements and poor management, which further exacerbated the damage caused by the water crises.

The main problems of China's water management system include:

- a fragmented approach to water management and the absence of coordination among government agencies;
- an inappropriate legal framework and lack of practicality; and
- inadequate participation of interested parties and insecurity of the rights and interests of the public.

Therefore, an institutional arrangement that enables proper coordination is critical to improve water governance. This would be strengthened by improved public participation.

It is a hard task to achieve the targets concerning water management put forward in the 11th Five-Year Plan. The targets for 2010 include: mandated targets, that is, to reduce the total discharge of chemical oxygen demand (COD) by 10 per cent and lower water consumption per unit industrial added value by 30 per cent; planned targets, that is, to raise the utilization coefficient of irrigation water up to an anticipated 0.5, and raise the percentage of urban sewage that is treated to a minimum of 70 per cent. But the indicators for 2000–2005 show that national sewage discharges are displaying an upward trend.[6] According to statistics from the Ministry of Construction, only 52 per cent of urban sewage was treated by primary or secondary treatments in 2005. Out of 661 cities,[7] 278 did not have sewage treatment plants (Sheng, 2006b). The sewage treatment plants in county towns are seen only in eastern regions, and in rural areas sewage is discharged directly into rivers without treatment. In addition, the pollution control tasks of major river basins were not completed. In 2006, the countrywide COD discharge increased by more than 2 per cent although the target was a 2 per cent reduction. Considering the economic growth trend, the complexity of the water problems and shortage of funding for pollution control, it is difficult to achieve the COD and urban sewage treatment targets. In other words, we have to overcome difficulties and obstacles in institutions, management, funding and technology in order to realize the above targets.

China's commitment and funding allocation are fundamental to achieving the mandated targets. Objectively, China's heavy industries are still developing and energy and raw material consumption are on the rise and far from reaching their peak. As a big, immature economic power, it is virtually impossible to arrest the growth trend of total pollutant generation over a short period of time. A large-scale, end-of-pipe pollution control strategy is essential to achieve the above targets. Such an approach is likely to reduce economic growth as a greater proportion of investment would go towards environmental protection. China must be soberly aware of this. It is estimated that in order to realize the goal of raising the percentage of urban sewage that is treated to 70 per cent by 2010, we have to increase investment by 332 billion RMB yuan (Sheng, 2006b), should economic growth be less than 7.5 per cent in the same period. However, economic growth in the first year of the 11th Five-Year Plan (i.e., in 2006) was over 10 per cent, which means a higher cost to achieve the target. Furthermore, the development of a circular economy and cleaner production is not an alternative to the above approach either. China's main objectives can only be to decouple material consumption and pollutant-generation growth from economic growth, raise water productivity and reduce the intensity of pollutant generation.

According to current Chinese law on water pollution prevention and control, the control of water pollutants is mainly applicable to water bodies that are unable to meet the standards for water environmental quality. The disaggregation of the current targets for cutting pollutants lacks a legal and strict scientific basis. On the other hand, only by making unified arrangements and planning reduced pollutant discharge from existing and new enterprises, raising the urban sewage treatment rate and reducing non-point pollution, is it possible to realize cost effectiveness (National Research Council, 2001). Theoretically, it is necessary to improve the environment and cut pollutant discharge. But rethinking is needed with regard to how much to cut and how and whether it is feasible, and long-term targets should be set on the basis of scientific analysis, rational institutional arrangements, supporting measures and good cooperation among different departments.

There is deviation in the orientation of investment in water management which urgently needs readjustment. According to the objectives set out in the 11th Five-Year Plan, the control of total pollutant discharge and water utilization efficiency are the core tasks with regard to water problems. But the investment, as shown in the past, has not been directed at water pollution prevention and control. Taking the seriously polluted Weihe River as an example, the water entering Baoji in Shaanxi Province is of Category II quality standard, but when it flows into the Yellow River, the quality is below Category V. However, in the Weihe River control programme developed in 2002, investment in pollution treatment accounted for only 10 per cent (Ma, 2007) of the total investment, while investment for water diversion accounts for 20 per cent. At present, all government departments are concerned about the safety of drinking water, but the measures adopted attend only to trifles and neglect the essentials, and fail to integrate the safety of water resources and their pollution control with river basin management and point source pollution control.

Furthermore, the investment is disoriented, failing to tackle the development of pollution, so water pollution is spreading towards the western part of the country, the rural areas and the upper reaches of rivers, while investment for pollution control is concentrated in the eastern part of the country, in cities and in the lower reaches of rivers. This has resulted in a lower rate of sewage treatment in urban areas and more serious industrial pollution in the western part of the country and in the upper reaches of rivers than in the eastern part. It is, therefore, pressing to adjust the orientation of investment and raise investment efficiency.

There is much potential for water efficiency improvement. In recent years, despite the fact that the central government has placed much importance on water conservation and water productivity[8] has gone up continuously (Figure 7.2), the water utilization rate has remained low and the situation of wasting water has remained very serious due to low water prices and other factors. The present annual irrigation water is about 360 billion m^3, extracted from rivers and groundwaters. As flood irrigation is still prevalent and canal leakage and evaporation are serious, the effective utilization is only 40–50 per cent as against 70–80 per cent in developed countries. The recycling rate of industrial water averaged only about 40 per cent as against the average of 75–85 per cent in developed countries (Ministry of Water Resources of China, 2003). The amount of water used per 10,000 yuan of industrial added value created is five to ten times that of developed countries. The water loss from leaks in supply pipelines in China's urban areas is about 20 per cent, three times as high as in developed countries (Qiu, 2005). Water shortages are serious in the western part of the country, including the arid and semi-arid areas of north-west China, but the water utilization rate in these areas is the lowest. Such low utilization rates make it hard to cope with the demands of economic development and the problem of reduced water resources on a per capita basis. But on the other hand, there is still a huge potential to save water both in industry and agriculture. There is also the possibility to raise domestic water utilization rates in urban areas.

Different policies should be adopted in different regions according to the different water problems and pressures. China's regional disparities are great.

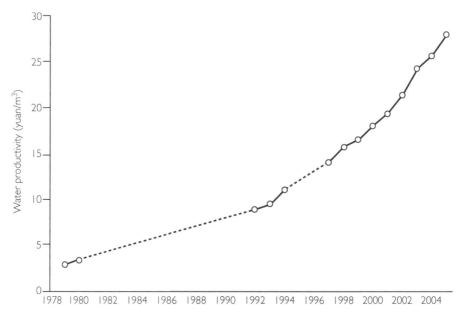

Figure 7.2 *Trends of water productivity (1979–2005)*

Different areas have different natural conditions and are at different stages of economic and social development. Their water problems and the pressure caused by the water problems they bear are also different. This determines that policies should be formulated according to the actual conditions of different areas and the policies must be flexible and adaptable. A survey shows that water resources in south-western China are better than in north-western China,[9] but the seasonal water supply is more unreliable and the use of unsafe water is higher in south-western China than in north-western China (Social and Sustainable Development Study Department of the National Research Center for Science and Technology for Development, FAFO Institute for Applied International Studies, 2006). The real problems of regional differences have raised new demands and challenges to water management departments in formulating policies.

Regional and river basin water problems are the result of long-term accumulation and so will be a long-term process for control and restoration (Wang et al, 2007). Although about 20 billion yuan have been spent over the past ten years in controlling water pollution in the Huaihe River, the pollution has remained serious. The water quality of one-third of cross-sections monitored is still below Category V. This indicates that the regional ecological and environmental problems require systematic, unified, coordinated and integrated management. In addition, because funds could not be put in place, only 70 per cent of the control projects were completed in the Haihe, Liaohe and Huaihe rivers and only half of the projects were completed in the Chaohu and Dianchi lakes in the 10th

Five-Year Plan period. And even if sewage treatment plants have been completed, they are unable to operate at full capacity because the pipelines and networks are not well matched, operational expenses cannot be ensured and fee collection policy has not been put in place. These problems have still not all been resolved effectively.

So, river basin pollution control and ecological restoration require sound policy and long-term unremitting effort. If we fail to adopt strong measures now to remedy the situation of inadequate policy and investment, the regional and basin-wide water problems in China will become more and more serious. Other river basins are likely to suffer the same fate as the Huaihe and Weihe rivers.

Experience of water governance and common understanding of an integrated approach to solve the water crisis

It is important to incorporate water-related objectives into the development strategy, plans and policies of the state and government departments. To take up the new challenge posed by water governance transition, it is necessary to develop a set of institutional arrangements and a governance system comprising multiple interest groups participating in public affairs on the basis of finding out the fundamental defects of the traditional ways. The basic principle is to seek harmony between man and nature, with equal consideration given to fairness and efficiency, and prevention and control so as to realize sustainable development (Qian, 2006). At the same time, it is necessary to set up a unified management system to coordinate cooperation among different interest groups and employ legal, administrative, economic and technical means to raise the utilization efficiency of water resources and effectively reduce pollutant discharge, safeguard ecological safety, minimize the risk of disasters and seek win–win situations for both the environment and development. For this purpose, we must make unremitting efforts to develop and execute all kinds of new ways of water governance. China is also accelerating changes in water resource management by advancing new concepts, such as promoting the integrated management of water resources, seeking harmonious relations between man and water, and increasing the environmental flows.

More and more countries and regions have begun to exercise integrated river basin management and integrated water resources management. Water problems are associated with the problems of environment and development. The 'Plan of Implementation of the World Summit on Sustainable Development' adopted at the 2002 UN Sustainable Development Conference called for support for developing countries to develop integrated water resources management and water efficiency plans by 2005. Through years of practice, people have become increasingly aware of the importance of basin-based integrated water resources management. In recent years, many countries have modified laws and regulations to encourage basin-oriented integrated management. The EU, for instance, adopted the EU Water Framework Directive in 2000, demanding that the then 29 member countries and neighbouring states formulate river basin management plans. South Africa adopted a water law in 1998, designed to encourage basin-based water resources management. New Zealand has even adjusted administrative boundaries based on river basin borders for the purpose of encouraging local government to exercise river basin management. The 12th session of the UN Sustainable Development Commission appealed to all governments to adopt measures for river basin management. China started the revision of general plans for all river basins in 2007 and will introduce the concepts of integrated management. All these examples show that water governance has entered a stage of integrated management.

Integrated management with supporting measures is the common path followed by all countries to improve water management and raise water management efficiency (Integrated River Basin Management Task Force of the China Council for International Cooperation on Environment and Development, 2005). First of all, making laws with regard to water resources and the water environment is of paramount importance in integrated management. Laws must establish objectives, principles, systems and operational mechanisms for river basin management and empower management organizations. Second, the enforcement of laws and regulations is a key issue. In order to empower the supervision and improve enforcement, procedural legislation, public participation and penalties for nonfeasance should be included in national law and regulations. Third, it is necessary to have an integrated programme with legally binding objectives and indicators to carry out the integrated management of river basins and water resources. Almost all water management organizations or river basin organizations have taken the mapping of an integrated programme as the most important part of their work, and it is through such programmes that they provide guidance with regard to water resources, the water environment and local river basin management. Finally, introducing economic instruments is an important way to promote behaviour changes. It is therefore necessary to increase supply, reduce discharge, economize on the use of water, encourage innovation and, through the fair distribution of water rights, reformation of water prices and creation of new financial and taxation policies, integrate organically the ecological compensation, technical progress and participation of related interest groups so as to provide adequate, safe and reasonably priced water to all people and improve services in the area of ecological restoration.

Conclusions and policy recommendations

Development objectives of water management in China from 2000 to 2020 have three dimensions:

1 To raise the utilization rate of water resources and lower the intensity of pollutant generation, which means raising the utilization rate of water resources by saving water, developing integrated utilization and cyclical use of water and, at the same time, changing the mode of growth, adjusting the industrial structure, developing the circulating economy (especially cleaner production) and technology, and lowering the pollutant generation per unit GDP.

2 To put the pollution control problems in the first place among all water issues, on the one hand, China should strengthen institutional arrangements, investment guarantee and law enforcement and supervision for controlling point source pollution. On the other hand, China should appraise the total load of river basin pollution, and then disaggregate it into different sources of pollution, including industrial, municipal and agricultural sources in order to strive to cut the total discharge of pollutants and improve water quality at basin level. From a long-term point of view, it is necessary to verify the total discharge load of pollutants of different river basins and water bodies used for different purposes and identify the specific pollution sources at different industrial points, urban sewage treatment plants and the rural non-point pollution sources. Only by doing so, is it possible to expect to integrate pollutant discharge control with the improvement of the environmental quality of water bodies.

3 To make programmes for improving freshwater ecosystems part of the targets for river basin planning and management. It is necessary to demarcate main rivers and river sections where development zones are banned or restricted according to the state plan for main functional zones and freshwater ecosystem requirements, so as to provide effective protection to river source areas, major water source areas, fishing zones and areas where rare and endangered species of aquatic wildlife are densely populated, natural wetlands and areas of estuaries. A directory should be compiled of rivers and river sections for priority protection that demarcates rivers and river sections with rich biodiversity, natural and cultural values, where water power development should be banned or restricted.

China's priority options and related policies for realizing the objectives of cutting pollutant discharge for the 11th Five-Year Plan are as follows:

• To give priority to point source pollution control. This is the most practicable and most mature means of controlling pollution at present. A pollutant discharge licence system and a pollutant discharge right trading system may be introduced according to related laws and decrees so as to exercise total pollutant control of industrial pollution points on the basis of river basin integrated control plans and, through promoting clean production, reduce the pollutant generation per unit of industrial added value.

- To make urban sewage control a target for reducing sewage discharge in the 11th Five-Year Plan period. In view of the requirements of economic development and the difficulty in reducing pollutant discharge of growing enterprises and the problem of pollutant transfer, it is an inevitable option to make the control of sewage, which accounts for about 60 per cent of COD load, the focus for total reduction. Fee collections should be reformed and a main fund channel set up that integrates government finances at all levels with public fee payments and, through the market-oriented reform of water affairs, promote and open a diversified investment and financing channel. It is also necessary to make unified arrangements in the handling of relations between urban sewage treatment and river basin integrated management of water pollution and strengthen the regulation by government.
- To introduce a new system of total control based on the maximum pollution load (environmental capacity). China may borrow the total maximum daily load (TMDL) method of the US in order to restore the environmental quality of polluted water bodies and formulate standards for discharge load according to the impact of point sources and area pollution sources on environmental quality. This method has been proved effective, but in China the process will have to start from the beginning. Basic data collection will have to be carried out in selected areas and the water body quality monitored and evaluated according to region and size of river basin, with the aim of creating a simple but effective model to calculate the pollution load and accumulate experience to apply the method elsewhere.

China aims to achieve water management objectives through the transition of the water governance model. China's water governance model has already started to, and will continue to, experience profound changes in the following aspects:

- First, innovative institutional arrangements are the basis of the new water governance model. Such an institutional framework would include a sound legal system, enforcement mechanism and incentive system.
- Second, water resources management, water environment management and freshwater ecosystems are integral to a unified water management system. Making a balance between centralization and decentralization, and a sound consultation mechanism would encourage integrated management and good governance. From a long-term perspective, an improved constitutional democratic system and market economy would further strengthen decentralization.
- Third, integrated water management is supported by legal, administrative, economic and technical strategies, in particular, the introduction of water pricing and water rights under a framework of an effective regulatory system and equitable institutional arrangements.

- Fourth, an incremental approach is required to achieve such a transition. It is a learning process and there is no established model. The Chinese have to be conscious of the inertia of the old system and look for options to minimize any negative impact and costs of such changes in order to achieve a smooth transition.

China aims to advance by stages and gradually set up a unified governmental integrated water management system. First, it is necessary to fully display the functions and powers of the existing water-related management departments, identify the priority areas of work of various departments and improve the coordination mechanism among different departments. Second, it is necessary to set up a steering organization to cover resources and the environment under the state council, such as the reinstitution of the former State Commission for Environmental Protection, with the leading member of the state council in charge of the area to become chairman. The commission will coordinate trans-sectoral, trans-regional or basin-wide major resource and environment problems. In the medium term, it is necessary to unify the functions of the government in the management of resources and environment, merge the resource and environmental protection functions of the existing water sources, environmental protection and forestry departments and form a unified department in charge of state resources and environmental protection to put resource- and environment-related affairs, including river basin-wide management, under unified control, coordinate steps of different departments and regions in major problems concerning river basin planning, standards and policies, and carry out integrated law enforcement and supervision. In the long term, it is necessary to decentralize the decision-making process, giving the local river basin organizations responsibility for making policy decisions on affairs associated with river basin-wide and regional water problems. The state department for resources and environmental protection should be responsible for overall planning, examination and approval, standards formulation, and regulation and law enforcement.

China aims to realize management transition and innovation, and employ all kinds of new management methods to meet the requirements of comprehensive management of water between 2000 and 2020. For this purpose, the Chinese should focus on management innovation in the following areas:

- Resource and environment performance management. To strive to lower water resource and energy consumption and water pollution generation per unit GDP through rational industrial layout, total pollution amount control and discharge permits, the assessment and rating of resource and environment performance, and encouragement of technical innovation in order to raise the productivity of water (Sustainable Development Strategy Study Group of the Chinese Academy of Sciences, 2006).
- Integrated river basin management. At the river basin level, the best way of changing the single objective control measure is to exercise integrated river

basin management, taking a river basin as a complete independent ecological system, and realize integrated water resources development, protection and management with sustainable development as the goal.

- Demand management. The transition from water supply management to demand management marks the maturity of a market economy. Demand management will also effectively check the orientation towards irrational supply expansion which is important for resource-short China. Water right management and water pricing reform are key and are an effective way to combine demand management with command-and-control measures.
- Adaptive management. In view of rapidly changing water pollution and the difficulty in realizing the set objectives, adaptive management is recommended, using dynamic and multiple objectives to direct planning and adjust the programmes of action according to the progress of development.
- Integrated risk management or crisis management. In order to cope with and adapt to uncertain water-related disasters, the threat of climate change and emergent environmental events, it is necessary to set up contingency plans on the basis of integrated risk management, integrating crisis management with routine management so as to effectively reduce the costs of disaster prevention and control, and contingency treatment.

China aims to continue water pricing reform with the aim of shaping a rational price mechanism. The goal of water right reform is to provide all people with adequate, safe and affordable water. Central to realizing the goal is a reasonable water pricing regime. Water pricing reform is the most important power behind the exercise of integrated management of water resources and environment, and is a way to raise the utilization rate of water, increase supply, reduce pollutant discharge and promote technical progress. It is one of the basic requirements for building a water-efficient society. Water prices should be composed of rates covering the cost of water resources, engineering infrastructure and environment. The future water pricing reform should be changed from rates covering only engineering costs to 'all-cost inclusive pricing', that is, it should include both water resource fees and water processing fees. The reform should follow the principle of 'compensating for cost, reasonable profits, higher prices for higher quality of water and fair burdens' in establishing a reasonable water rate formation mechanism, and set up a hearing system for water prices so as to create a public decision-taking mechanism and price supervision mechanism in water rate management. Considering the affordability, such reform should start in cities, actively introducing the 'step water prices', and should give full consideration to regional differences. The reform of the rates for water used in agriculture should focus on taking stock of the practice of 'hitchhiking' in raising prices, putting to order the water rate collection in irrigation areas and improving the measuring facility.

China aims to establish a diversified resource and environmental protection financing and investment mechanism. The realization of the objectives of reducing major pollutant discharge depends on the placement of investment in environmental protection. It is essential to smooth out the existing investment channels and open new ones. First, it is necessary to display the role of environmental finance as the main channel, continue to invest funds by selling treasury bonds, and strengthen budgetary investment in solving trans-regional pollution problems. It is also important to give priority to solving the problem of trans-regional river basin pollution control and use the funds according to unified arrangements for a whole river basin. Second, while carrying on with water pricing reform, it is necessary to improve the existing mechanism for collecting fees for pollution treatment and develop guidance with regard to the standards for collecting fees for sewage treatment, so as to form a reasonable price formation mechanism. Third, it is important to set up an environmental capacity compensation system and, through pilot projects, gradually set up a TMDL-based total control system and establish the initial pollutant discharge right that must be bought and, at the same time, set up a pollutant discharge right trading system. Fourth, accelerate the pace of establishing a market in environmental infrastructure facilities to break the government's monopoly on public utilities and introduce competition, accelerate the pace of system conversion of urban water affairs units, adopt a variety of franchised management models to encourage idle social funds and foreign investors to participate in environment-related infrastructure construction, develop all kinds of financial instruments, especially reform-associated laws and regulations, and operate municipal bonds for priority use in building municipal and environmental protection facilities.

China aims to raise innovative abilities in science and technology and set up a science and technology supporting system including both hardware and software aspects for solving water problems. It is necessary to find scientific ways and technical means of solving problems in priority areas by fully analysing the water problems confronting China, strengthen original innovation, integrated innovation and broad international cooperation, while paying attention to research and development and the application of appropriate technologies. The science and technology supporting system for solving water problems should include:

- theories and technologies associated with water resources and water pollution management, such as water resources management, river basin master planning, water rights and pollutant discharge right distribution, water pricing, standards for resource and environment performance by industry, dynamic monitoring and management of information;
- utilization rate of water resources, alternative water resources and projects, technology, materials, products and equipment (such as desalination of sea water and reuse of grey water);
- all kinds of sewage treatment technology, integrated regional pollutant control, safe drinking water technology, ecological restoration technology and technologies associated with environmental health; and

- technologies for preventing and controlling water-related disasters, including disaster precautions, forecasting, emergency treatment, disaster relief and related management technologies.

The development and application of the above technologies will promote the building of a resource-efficient, environment-friendly and water-efficient society.

Notes

1　Guan Zi is a famous politician who lived 2000 years ago.
2　The concept of the hydraulic society is a system of centralization of state power through exclusive control over water such as building large-scale flood control and irrigation engineering in an agrarian society.
3　Category V of the State Environmental Quality Standard for Surface Water (GB3838-2002) means that the water is suitable for agriculture and general use.
4　The traditional water management system in China features highly centralized decision making and separate sectoral management with no effective coordination at central level.
5　Water governance in China is no longer concerned solely with the conquering or controlling of water by engineering measures in the traditional sense. Instead, it respects nature and employs all sorts of means to manage water well so as to realize harmony between man and water. Besides, water resource management in China usually means management of water quantity and is mainly the responsibility of the Ministry of Water Resources (MWR).
6　The total national sewage discharge was 41.5 billion metric tonnes in 2004 and 52.4 billion metric tonnes in 2005, growing by 26.4 per cent from 2000 to 2005, according to the statistics from the State Environmental Protection Administration (SEPA); the figure was 62 billion metric tonnes in 2000, 71.7 billion metric tonnes in 2005, growing by 15.6 per cent from 2000 to 2005, according to the statistics from the MWR. The data come from *Annual Statistic Report on Environment in China 2005* (SEPA of China, 2006) and the *Water Resources Bulletins* of the MWR for all years.
7　The city is set administratively by governments, not including all towns at county level. Among 661 cities, 4 municipalities were directly under the central government, 283 at prefecture level and 374 at county level.
8　Water productivity refers to the value of economic goods and services per cubic metre of water extracted from the natural environment and is expressed as GDP/water use.
9　Due to reasons of statistical method, there may be errors in these data. But in general, compared with developed countries, there is no doubt that the water utilization rate in China is low.

References

Ge, J. (2005) 'River ethics and continuation of human civilization', Wenhui Bae, 7 February 2005, www.tecn.cn/data/detail.php?id=5865
Gu, H. (ed) (2006) *China's Water Governance in Historical Perspective*, 2nd edn, China Water Resources and Hydropower Publishing House, Beijing
Integrated River Basin Management Task Force of the China Council for International

Cooperation on Environment and Development (2005) 'Promoting integrated river basin management and restoring the living rivers', in State Environmental Protection Administration (SEPA) and CCICED (eds), *Proceedings: The 3rd Meeting of the 3rd Phase of CCICED*, Beijing, pp213–229

Ma, Z. (2007) *Evaluation of the Implementation of Water Pollution Prevention and Control Plans in China: The Case of Huai River Basin*, www.chinaeot.net/wb_wateraaa/Report%20-%20Huai%20river%20water%20pollution%20plan%20FINAL%20EN.doc

Ministry of Water Resources of China (2003) 'Problems and countermeasures with regard to China's water issues', *Water Resources Newsletter*, no 21–23, www.mwr.gov.cn/zwxx/20030305/1183.asp

National Research Council (2001) *Assessing the TMDL Approach to Water Quality Management*, National Academy Press, Washington, DC

Qian, Z. (2006) 'Harmony between man and nature', speech at the Conference of Academicians of the Chinese Academy of Sciences and the Chinese Academy of Engineering, Beijing

Qiu, B. (2005) *Situation, Challenges and Countermeasures in Urban Water Issues*, Speech at the First International Symposium of China's Urban Water Affairs Development Strategy, www.csjs.gov.cn/sys/FirstPage_detail.aspx?TabaleName=tmp2&id=2737

SEPA of China (2006) *Annual Statistic Report on Environment in China 2005*, China Environmental Sciences Press, Beijing

Sheng, H. (2006a) 'Report of the Law-Enforcement Examination Group of the Standing Committee of the National People's Congress on the Implementation of the Water Pollution Prevention and Governance Law of the People's Republic of China', speech at the 16th session of the 10th NPC Standing Committee (29 June 2005), in The First Bureau of Secretariat of the Standing Committee of the National People's Congress (eds), *Collection of Reports on the Examination of Enforcement of the Water Pollution Prevention and Control Law of the People's Republic of China by the National People's Congress 2005*, China Democracy and Law Publishing House, Beijing

Sheng, H. (2006b) 'Report on tracing implementation of the environmental protection law by the law enforcement examination group of the Standing Committee of the National People's Congress', Speech at the 23rd meeting of the Standing Committee of the 10th National People's Congress, 26 August 2006

Social and Sustainable Development Study Department of the National Research Center for Science and Technology for Development, FAFO Institute for Applied International Studies (eds) (2006) *Life in Western China: Tabulation Report of Monitoring on Social and Economic Development of Western China*, China Statistical Press, Beijing

Sustainable Development Strategy Study Group of the Chinese Academy of Sciences (2006) *China Sustainable Development Strategy Report 2006 – Building a Resource-Efficient and Environment-Friendly Society*, Science Press, Beijing

Sustainable Development Strategy Study Group of the Chinese Academy of Sciences (2007) *China Sustainable Development Strategy Report 2007 – Water: Governance and Innovation*, Science Press, Beijing

The Editorial Board of China's National Assessment Report on Climate Change (2007) *China's National Assessment Report on Climate Change*, Science Press, Beijing

Wang, Y. (2002) 'Environment, development and governance: How to face environmental challenges in the new century', in CCICED/SEPA (eds), *Proceedings: The First Meeting of the 3rd Phase of CCICED*, Huawen Press, Beijing, pp188–207

Wang, S. (2003) *Resource-oriented Water Management – Towards Harmonious Coexistence between Man and Nature*, China Water Resources and Hydropower Publishing House, Beijing

Wang, Y. (2006) 'China's environment and development issues in transition', *Social Research*, vol 73, no 1, pp277–291

Wang, Y., Li, L., Wang, X., Yu, X. and Wang, Y. (2007) *Taking Stock of Integrated River Basin Management in China*, Sciences Press, Beijing

Wittfogel, K. A. (1957) *Oriental Despotism: A Comparative Study of Total Power*, Yale University Press, New Haven, Connecticut

Part II

Irrigation Technology to Achieve Water Conservation

8

Irrigation Technology and Water Conservation in Jordan

Munther J. Haddadin

Introduction

The Hashemite Kingdom of Jordan is a young country in the Middle East, independent since 1946 with historic ties to the UK and, since 1957, with the US. It has witnessed waves of incoming refugees the first of which were Palestinians in 1947–1948 in the wake of the UN Partition Resolution of Palestine and the first Arab–Israeli war. The second was between 1950 and 1967 when Palestinians from the West Bank shifted residence to Jordan after the two entities became integral parts of the Hashemite Kingdom, and the third was also Palestinian during and in the aftermath of the June 1967 war between Israel and the surrounding Arab countries. More waves ensued in the 1980s and thereafter, this time from the east as Iraq forcefully occupied Kuwait in 1990 and the succeeding wars thereafter.

The above is not presented by way of historical account, but to show that while the natural resources of the country, including water, are finite, the population growth defied biological normal growth and increased at about double the biological rates. The population of the Kingdom has increased from about 350,000 in 1946 to 5.6 million people today.

Jordan's climate belongs to the arid and semi-arid environments. The majority of its territories, or about 90 per cent, receive less than 20mm of rain per year. The total annual precipitation over the country's territories averages 8200mcm (million cubic metres) of which 6582mcm (80 per cent) evaporates, indigenous surface water averages 554mcm, 198mcm forms indigenous renewable groundwater resources and 866mcm is green water that supports rain-fed agriculture and pastures for livestock feed. Exogenous water consists of 68mcm subsurface flows from Syria, 60mcm surface water transfers from Israel, and 245mcm surface flows from the Yarmouk river (Haddadin, 2006). The above indicates that the total renewable freshwater resources average 1991mcm/yr. With a population of 5.6 million, the per capita share (as at 2007) is 355m^3. About 66m^3 of that is exogenous and can be affected by the actions of neighbouring Syria and Israel.

Jordan falls in the lower middle income category.[1] Its annual per capita need for water is 1700m^3 (Haddadin, 2007). Therefore, the water availability of 355m^3 is only 21 per cent of its needs which creates a huge strain on water resources in the country. Such strain imposes unusual stress on water resources management: over-abstraction from aquifers has been widespread. In 2007, an average of 200mcm was over-abstracted from renewable aquifers, 85mcm from fossil non-renewable aquifers, and treated wastewater reuse has been added to the renewable water resources at an average of about 90mcm.

The pattern of water use in 2002 was: 250mcm for municipal purposes, 37mcm for industrial purposes and 1377mcm for agricultural purposes. This last figure includes 80 per cent of the green water potential or 675mcm, 523mcm of blue water including 167mcm of over-abstraction from groundwater aquifers, 72mcm of treated wastewater effluent and 89mcm of fossil water use. This pattern shows that of the total water used in 2002, 70 per cent of the fluid water was used in irrigation and 83 per cent of the total water resources (including green water) was used by agriculture.

The above shows the share that irrigation took of the fluid resources. Improvement of the overall irrigation efficiency would translate into savings of sizeable water flows, thereby showing the importance of irrigation technology for Jordan and the water-strained countries.

Irrigation technology in Jordan

Irrigated agriculture in Jordan has been practised since the dawn of human civilization. Human settlements dating back 11,000 years were uncovered in the Yarmouk Valley. Pella in the Jordan Valley was once a coastal town. Irrigation systems dating back to the Nabateans (300BC) have been found south of the Dead Sea and clay-piped conveyance systems were known then. Roman irrigation systems were discovered too. In the Jordan Valley there is a score of human settlements dating back to various historical eras. However, the Jordan Rift Valley was populated and de-populated throughout its history denying it the continuity in irrigated agricultural production; something that the Nile Valley, for example, enjoyed. The interruption of irrigation did not allow the advancement in irrigation technology through a process of development.

Surface irrigation has been the technology in vogue in recent centuries. Surface earth canals were built to convey the water from source to destination. Surface irrigation methods were used to distribute the water in the fields. More often than not, the fields were levelled to guarantee laminar uninterrupted flows in the furrows and smooth application of water to the plants. No soil salinity was observed as a result of land levelling.

With the advent of cement, concrete surface canals were introduced in Jordan in the mid-1950s. The concrete lining was justified on the grounds of saving much-needed irrigation water consisting of base flows of streams. The introduction of concrete lining was through a pioneer irrigation project in the Middle Jordan Valley where an agricultural research and extension station was established.

The government project was emulated on a small scale by farmers who developed their irrigated agriculture by withdrawing water from perennial tributaries to the Jordan River, known as 'side wadis'. The water savings allowed expansion of irrigated areas while the water share from the stream, per farmer, remained unchanged. Agricultural cooperatives were soon established and borrowed from specialized credit corporations, mainly the Government's Agricultural Credit Corporation, to improve their irrigation water conveyance. The on-farm irrigation technology remained the historic methods of basin, furrow or flood irrigation techniques, and the agricultural yields were compatible with what was known as the yields of surface irrigation methods.

Advancement through cooperation

The first official shift in the adoption of irrigation technology came in 1958 when the US, acting through the Technical Cooperation Agency, the predecessor of the US Agency for International Development, advanced a grant to Jordan to finance the first stage of the East Ghor Canal Project. The project consisted of a main concrete lined canal with a carrying capacity of $10m^3/s$, and a distribution network of concrete lined canals, check structures, siphons and aqueducts, and farm turnouts. A special legislation was enacted for the project which allowed land expropriation and parcelling into farm units ignoring the previous land borders, and redistribution to landowners and to landless farmers. This was a successful pattern of land reform in which the large and small landowners and landless farmers benefited. The landowners benefited by having their lands irrigated by the project without which such lands would not be productive under rain-fed conditions, and the landless farmers by owning irrigated land. Financial accounting was guaranteed whereby the owners who lost land would be compensated and the farmers who received land would be charged.

Land levelling was part of the project. Lands that fell above the water line were left unirrigated because pumping was not a feasible option in the 1960s. Agricultural research and extension were strengthened and the irrigation project was a big success. The US funding was sufficient to irrigate 11,500 hectares (ha) of arable land in the East Jordan Valley and that was consistent with the water shares defined for Jordan by a US-sponsored water division formula to co-riparians in the Jordan basin – Lebanon, Syria, Jordan (including the West Bank) and Israel. The motives behind the US funding were to consolidate the water sharing formula, to create job opportunities in Jordan which was overburdened with the influx of Palestinian refugees, and to help settle them in the Jordan Valley (Haddadin, 2001). Jordan benefited from the production in the Jordan Valley, almost a natural hothouse, of off-season fruits and vegetables both in the form of supplies for domestic consumption and for export, from the jobs the project created and the immense positive social and environmental impact the project had on the country during construction (1959–1966) and thereafter (Haddadin, 2006).

Resumption of agricultural activities

Agricultural activities in the Jordan Valley all but halted in the wake of the June 1967 war between Israel and the surrounding Arab countries. The valley became the scene of paramilitary activities when the PLO took positions there from which they attacked Israel and the latter retaliated. The valley population defected to the highlands for safer refuge and the irrigation project ran at a low rate. This situation lasted from 1967 to 1970 when the Jordan Arab Legion clashed with the PLO fighters and evacuated them by force in 1971. The re-establishment of law and order encouraged the population to return to the valley and by October 1972 the population level had rebounded to its pre-war level of 64,000 people.

The Jordanian government prepared a three-year development plan for the Jordan Valley (1972–1975) and announced its intention to focus on the valley's integrated social and economic development with irrigated agriculture as its backbone. The government's intentions encouraged farmers to reactivate their farming efforts and to invest more in their farms. It was at this time that innovative irrigation methods found their way to the Jordan Valley through private entrepreneurs.

The private entrepreneurs were mostly Lebanese businessmen who had seen the drip irrigation methods in Australia and imported them into the United Arab Emirates, a Gulf water-poor country. From there, the same entrepreneurs brought the drip irrigation method to the Jordan Valley and promoted its use with selected farmers. The drip plastic pipes and the emitters were brought in from Cyprus before they were manufactured in Jordan.

Almost concurrently, protected agriculture was also introduced. Plastic tunnels were tried by some innovative farmers who used reinforcing steel bars (8mm diameter) available on the market to make arched supports for plastic tunnels over vegetable seedlings. Surface irrigation methods were used to irrigate the plastic tunnels at the beginning. Even with this primitive irrigation method, the agricultural yield increased almost two-fold or more. By 1974 drip irrigation methods and plastic houses as agricultural production practices had not begun in Jordan.

Official introduction of advanced irrigation methods

The Jordan Valley development plan called for an expansion in irrigated agriculture by about 72 per cent, from 11,500 to 19,800ha with water drawn from the uncontrolled flow of the Yarmouk river, which Jordan shares with Syria upstream and Israel downstream, and on the flow of side wadis, the most prominent of which is the Zarqa river whose flow would be regulated to augment the Yarmouk, especially in the dry months. One of the irrigation expansion projects was an extension of the East Ghor Main Canal by 18km which would add 3650ha to the irrigated area. The project was designed just like its predecessors, i.e., concrete lined canals and check structures. The project was in the bidding stage when a mission from the International Bank for Reconstruction and Development (IBRD) was visiting Jordan to appraise yet another irrigation project in 1973.

The leader of the IBRD mission was a professional irrigation engineer, the late Dr Guy Lemoigne, who met with the president of the Jordan Valley Authority, Omar Dokhgan, and convinced him to adopt piped distribution systems and a sprinkler method of irrigation. Convinced of the transformation on the grounds of better overall irrigation efficiency, Dokhgan cancelled the canal-extension bid whose documents were already distributed to contractors, and instructed the consultants to redesign the project adopting piped distribution systems. Three other projects of various sizes were also designed adopting piped distribution systems. Furthermore, 1000ha of already irrigated area (using the surface canal distribution network) was converted to a piped irrigation network.

The US Agency for International Development (USAID) was approached to support a sprinkler irrigation system on the farms served with the piped distribution network in which the pressure was generated either by gravity to low-lying farms or by supplementary pumping to farms close to the East Ghor Main Canal. The pressure at the farm turnout was designed to be three atmospheres, enough to operate sprinklers on the farms. The USAID advanced a loan of US$4 million to procure sprinkler sets from the US. By the time the irrigation projects were completed and ready for operation in 1979, the sprinkler equipment was procured and stocked in Jordan, ready for distribution to interested farmers on credit advanced by the government's Agricultural Credit Corporation (ACC).

By that time, the drip irrigation systems and the protected agricultural practices had gained momentum in the Jordan Valley. Additionally, farmers were not attracted to sprinkler irrigation methods to irrigate vegetables or trees for fear of creating an environment conducive to the growth of pests. However, sprinklers were adopted to irrigate field crops such as wheat and barley which were planted in the valley not because they were more profitable than vegetables, but because they were part of the crop rotation farmers liked to pursue.

The result of competition between the drip methods and the sprinkler methods was decisively in favour of the drip methods, and the sprinkler equipment was sold as raw aluminium material or to farmers on the Jordanian Plateau who irrigated field crops.

Factors promoting the advanced irrigation technology

Several factors influenced the adoption by farmers of advanced irrigation methods and also prompted the Jordan Valley Authority to support the trend and urge the ACC to advance credits to interested farmers. Such factors are listed below.

Erosion of water shares

Jordan's share in the Yarmouk was progressively diminished by Syrian abstractions upstream. The diversion of Yarmouk spring–summer flows to Jordan dropped from about $4.8 \text{m}^3/\text{s}$ to about $1.2 \text{m}^3/\text{s}$ between 1963 and 2006.

When the expansion in irrigated agriculture was decided in 1972, the water diversions from the Yarmouk and the contributions from the side wadis and the King Talal Dam (then under construction) on the Zarqa River were sufficient to meet the water requirements of the expanded irrigation in the Jordan Valley. However, as the expansion projects were implemented, Syrian withdrawals from the Yarmouk were increasing at the expense of Jordan's diversions from the river. This prompted the Jordanian government to convert the surface canal distribution networks to pressure pipe networks to save water by increasing the efficiency of distribution, thus making up for the shortages imposed by the additional Syrian withdrawals. The farmers responded in kind and invested in advanced irrigation technology using mostly drip irrigation methods. The farmers, led by the Jordan Valley Authority, organized themselves into water user groups and took charge of water distribution along selected irrigation laterals; the process is continuing (Salman et al, 2008).

The end result has been a partial make-up for the losses caused by Syrian abstractions from surface water through dams it had built in the Yarmouk catchment and the over-use of springs, in addition to over-abstraction from the groundwater aquifer feeding lower springs in the basin which were among

Jordan's water share (Jordan Valley Authority, 1987). Water shortage, therefore, was a major reason for the propagation of pipe irrigation networks and advanced irrigation systems.

Increase in irrigation efficiency

Advanced irrigation technology eliminates most of the handicaps associated with traditional irrigation methods. An important outcome of advanced irrigation technology is the saving of water that would otherwise be lost to evaporation and percolation. This increases irrigation efficiency and reduces the need for subsoil drainage. In Jordan, in dry seasons with water shortage, an irrigation efficiency of 82 per cent was recorded. This was the result of adopting drip irrigation methods and piped water distribution networks.

Increase in agricultural yield

The adoption of advanced irrigation technology, in addition to water savings, increased the agricultural yields. It also facilitated the employment of advanced agricultural practices by which plastic mulch is used to cover strips in the field where seeds or seedlings are planted. Plastic tunnels or plastic houses are then used to protect the plants against cold fronts that can damage the crop and cause losses. The same plastic mulch is usually installed ahead of planting time to fumigate the soil. The increase in yield more than pays for the additional capital investment with a fair amount left for profit.

Improving the labour environment

Traditional on-farm surface irrigation methods entail harder labour efforts to make and maintain the irrigation ditches and basins. Advanced irrigation technology eases the labour effort and promotes labour productivity.

Promoting treated wastewater reuse

Risks associated with treated wastewater reuse are reduced by the use of advanced irrigation technology associated with plastic mulch. Contact between the workers on, and visitors to, the farm and irrigation water is almost all but eliminated thus protecting these individuals against possible diseases caused or transmitted by contaminated irrigation water. Moreover, contact between the fruits and the irrigation water is minimized thus protecting the public against health hazards caused by the treated wastewater.

Treated wastewater reuse brings advantages of plant nutrition from nutrients contained in the wastewater. This is an added benefit of the wastewater reuse.

Optimizing the water output

Advanced irrigation technology is open to the use of automated control systems for water application. For example, sensors attached to the stems of trees flag the need for water to a control centre which issues the order to have water applied to the set of trees around that sensor. This will boost on-farm irrigation efficiency and crop yield. The tonnage per unit flow of water can be maximized and water losses materially minimized. In the early stages of its introduction, advanced irrigation systems yielded about three times as much as traditional surface irrigation systems. And when protected agriculture was introduced using plastic houses and drip systems, the yield was boosted to ten times or more.

Improving environmental conditions

Advanced irrigation systems reduce the chances of weed growth on the farm. In addition, the use of plastic mulch, usually associated with drip systems, helps protect the soil against pesticides sprayed on the plants. And animals are protected because contaminated water flows in closed pipes and is not exposed in storage ponds where birds and wild and domesticated animals may drink. This problem was witnessed on some farms in the Jordan Valley where treated wastewater is used in irrigation. Advanced irrigation technology reduces such risks.

The offtake pipes from the main King Abdallah Canal in the Jordan Valley each incorporate a screen installed in a special groove. The screens prevent the inflow of suspended pollutants such as weeds, algae, pieces of wood, etc. into the pipes and ensure the absence of such suspended matter. De-silting basins are also installed at the points where the pipes come off the main canal so that the finer suspended matter can settle before the water is pumped to the distribution network or allowed to flow into it by gravity. Finally, sand filters are installed by farmers at the farm gate immediately after the farm turnout to ensure clarity of water. All these systems undergo a backwash service to ensure their proper operation. The water conveyance and distribution by the Jordan Valley Authority runs in closed pipes except for the main King Abdallah Canal which runs down the valley for a distance of 110.5km. The reservoirs of the storage dams in addition to the main canal are exposed to evaporation losses.

Because of the cycles of service of irrigation water by the Jordan Valley Authority (average of three days per week for six to eight hours per day), most farmers store irrigation water in ponds and pump the stored water to suit their own irrigation cycles on the farm. The surface of the ponds is exposed to the atmosphere and to evaporation losses.

The adoption of advanced irrigation systems reduces evaporation losses and also enables better control of water application at the plant zone. Under conditions of scarcity brought about by the aridity of the region and by man's intervention, the cumulative advantage of the advanced irrigation system allows the irrigated areas to be served with water, and increases the output per unit area. The labour input in the irrigation process was also reduced and stands to be decreased even further by the introduction of automated irrigation services.

Improving farm economics and quality of life

Farm economics are enhanced using advanced irrigation technology. Yields are increased, water is saved, labour productivity is enhanced and farm income improves. This creates positive social impacts on the farming community and improvements in living conditions. It was possible to boost the per capita share of income in the Jordan Valley, over 13 years of irrigation development using advanced irrigation systems, to match the average per capita income in the country (Shibley et al, 1987). The social impact of the integrated development was very pronounced. The quality of life improved, the cultural environment transformed, the role of women expanded to the provision of services, and the demand for agricultural labour dictated the importation of labour from nearby countries.

Irrigation technology on the plateau

The plateau of Jordan comprises cities, towns and villages and also has arable land cultivated through rain-fed agriculture. The irrigation benefits in the Jordan Valley as described above triggered irrigation practices on the plateau and in the Badia region that receives modest rainfall in the east. Groundwater is used to irrigate these lands.

Unlike the development of the Jordan Valley where the government invested to irrigate the arable lands, the development of irrigation in the plateau was carried out by the landowners. The government issues licenses to drill for, and permits to abstract, groundwater. Landowners pay for the drilling of wells to reach the aquifers (usually at depths between 150 and 350m) and they care for the on-farm development in a similar way to the Jordan Valley farmers.

The irrigation technology that was introduced in the Jordan Valley, mostly drip irrigation methods, quickly spread to the farmed areas on the plateau. Economic reasons, similar to the gains that accrued from agriculture in the Jordan Valley, prompted the spread of irrigation technology on the plateau. Water charges were also another motive. Each well abstracting groundwater is allowed a specified quantity of water for free to compensate for the development cost the farmer had incurred. Any quantity abstracted over and above the specified amount is charged for at a high price (MWI, 2002). Water savings here have substantial financial and economic value. In the Jordan Valley, the Jordan Valley Authority distributes set amounts of water, measured by the duration of service each time, as opposed to the plateau farmers who operate the water system themselves.

To ensure the application of the groundwater management regulations, the Water Authority of Jordan created a new directorate and staffed it with personnel borrowed from the Retired Military Personnel Corporation. The staff of the new directorate visit wells abstracting groundwater, enforce the installation of water meters, and make periodic readings to calculate the abstracted water as a function of time and impose charges where the abstraction exceeds the allowable free flow per annum. Owners of illegally drilled wells were given a grace period to legalize their wells in accordance with the new regulation and all the wells now have a proper record at the water authority. A campaign of confiscating unlicensed drilling rigs was mounted in 2002 and many of them were held in the authority's yards. Penalties were imposed by law and a much better control on well drilling is now practised.

In the southern Jordan Valley, violations continued, not only by trespassing on groundwater (by drilling groundwater wells without a licence and abstracting groundwater without permits) but also on land belonging to the Jordan Valley Authority or to the state or both. In 2006, when Jordan Valley Authority and Water Authority of Jordan personnel seemed unable to control the situation, the government sent in armoured police regiments and took care of all the violations by force.

The total irrigated area on the plateau today matches the irrigated area in the Jordan Valley, about 30,000ha each. Almost all the irrigated areas, with very few exceptions, use advanced irrigation technology. In addition to the water saving, there is also less need for agricultural drainage. An added benefit is the improvement of the quality of fruits grown under advanced irrigation methods, especially when protected agricultural production techniques are employed.

The spread of irrigation technology prompted the establishment of a new industry in the country to manufacture drip irrigation pipes and equipment, with distribution agencies spread all over the country. This created many jobs and shortened the maintenance response times because the supply sources of irrigation equipment became local.

Contemporary problems facing irrigated agriculture

Irrigated agriculture in Jordan is facing hardships emanating from the elimination of subsidies on consumer goods such as water, food items, fuel, etc. and on the escalating costs of farm inputs imported from the major trading partners in Europe. A third hardship originates in the lack of availability of unskilled agricultural labour. Another is the sustainability of irrigation water supplies. A perennial problem is agricultural marketing, and a sixth problem is the effect on irrigated agriculture of Jordan's membership of the World Trade Organization (WTO).

Elimination of Treasury subsidies on consumption

Perhaps the most important item in this category is the impact on the price of fuels and lubricants. The government decided to lift subsidies from fuel consumption in line with its programme of economic structural adjustment started in 1989. The prices jumped to match the free market prices which meant that those of gasoline and diesel doubled and the price of cooking gas tripled overnight. These costs are adjusted to reflect the free market prices every month.

As is well known, the hike in energy prices brings about a chain of price increases in most commodities, if not all of them. The purchasing power of Jordanian consumers diminished despite increases in wages to partially compensate for the increase in fuel prices. Electricity prices were raised as of 14 March 2008 (*Al Rai*, 2008) to reflect the increase in the fuel oil used in steam power plants and the price of gas used to drive gas turbines.

The next to follow will be the adjustment of the water tariff in the Jordan Valley to cater for the increase in pumping costs on the one hand and to eliminate the subsidies for irrigation water on the other. Although not yet decided, one can see it on the horizon.

Irrigated agriculture is and will be affected by these measures. Diesel oil is used extensively in irrigated agriculture to drive vehicles, tractors, pumps and engines. Electricity is used to drive pumps on many irrigated agriculture farms. The observed trends of increased prices of commodities, even locally produced inputs such as organic fertilizer, negatively affect agriculture. The cost of packing the produce has increased, almost doubled, because boxes are made of polystyrene, a petroleum-based material, and the price of drip irrigation pipes, also made from petroleum-based products, has risen.

Cost of agricultural inputs

Agricultural inputs of fertilizers, pesticides, machinery and spare parts have increased drastically. The popular brands of such inputs have their source and origin in Europe, the euro currency area. The euro has appreciated substantially with respect to the US dollar to which the local currency, the Jordan dinar, is tied (1JD is worth US$1.408, a rate that has held steady since the early 1990s). By mid-March 2008, the euro was worth over US$1.55, up from a low of US$0.86 in 2002.

The same problem exists with regard to imports from other major trading partners such as Japan. Agricultural motor vehicles are mostly of Japanese origin and the same is true of the yen with respect to the dollar/dinar as with the euro.

Agricultural labour availability

Direct agricultural labour has been heavily augmented by workers from Egypt and other developing countries like India, Pakistan and Sri Lanka. The

Government has issued new covenants to organize the inflow of agricultural labour and minimize the defection of agricultural foreign workers to other sectors of the economy. There is preferential treatment in licence fees in favour of agriculture that is not accorded to other sectors of the economy. As a result of these new covenants, foreign workers are not as easily admitted to Jordan as has been the case for the past 20 years. Additionally, the wages of these foreign workers have increased in the wake of the upward adjustment of fuel prices and the wave of price increases that accompanied it. The difficulty associated with the admission of foreign workers diminished their supply and, in the free market economy of Jordan, drove up the wages of available workers.

Sustainability of irrigation water

Irrigation water in the Jordan Valley is subject to natural seasonal variations in precipitation, but that was accounted for in the sizing of storage dams and in the operation of these dams. What was not accounted for is the level to which the upstream riparian party on the Yarmouk, namely Syria, has exploited the surface and groundwater resources of that basin. In 1966, the Yarmouk base flow was sufficient to support about 11,500ha but today it is hardly sufficient to support even half that area.

Another pressure on irrigation water resources is the need to curtail withdrawals from aquifers to bring the total withdrawal from them to sustainable rates. This means a reduction of abstraction by about 30 per cent of what is abstracted today. Since abstraction for domestic and industrial uses cannot be curtailed, then irrigated agriculture has to face the reduction.

A third source of pressure is the calls by some politicians and donors to divert more of the irrigation water to municipal and industrial uses. In the past, such policy was accompanied by the treatment and reuse of municipal waste-water, and this policy is likely to continue should more decisions be made to divert agricultural water to municipal uses.

Agricultural marketing

The current system of agricultural marketing relies heavily on local wholesale markets run by the respective municipalities against a fee which is a percentage of the sale price of goods sold. Even agricultural exports originate in these wholesale markets. A farmer who supplies his produce to the central market has to pay the costs of the containers/boxes, transport, the municipality percentage and the auctioneer. Of the net that he receives, he has to pay for farm inputs, labour and other costs. Unless his final net income is positive and is enough for him to maintain an acceptable standard of living for his family, he will obviously have to abandon the trade and look for some other occupation.

Export of fruits and vegetables is done by private exporters – some truck the exports to the states of the Gulf Cooperation Council and others airlift the exported goods to Europe albeit on a very modest scale. Several attempts have been made to have the farmers, through their Farmers' Association, compete in the marketing process but that did not succeed except partially in securing some farm inputs.

More recently, HM the King, during a visit to the northern Jordan Valley, pledged to have an airport built there to help the export of valley produce to Europe. Hopes are high on the provision of export marketing infrastructure and organization of the marketing process to do the producer more justice.

Membership of the WTO

Jordan's membership of the WTO allows the opening of its markets to other members after a 10-year grace period during which Jordan is supposed to adjust its conditions and get ready for competition in agricultural produce in its own markets. The prime agricultural products that Jordan brings to markets, both local and regional, are the off-season fruits and vegetables. Citrus fruits, for example, start hitting the market by late October; vegetables start arriving at markets in November. There is hardly serious competition to the Jordan Valley produce in winter except from temperate zones.

But there are alternatives to citrus fruits, for example, in the form of other fruits that can be imported into the Jordanian market such as apples, bananas and pears. These have the effect of reducing the demand for Jordanian fruits in winter and the reduction of market prices. Little has been done to prepare the Jordanian agricultural market for the competition imposed by imports.

Promotion of advanced irrigation systems

There has been widespread adoption of advanced irrigation systems by farmers across Jordan and this has provided an opportunity for industries to emerge. Manufacturers of polyethylene pipes of various diameters and strength and their fittings, drippers, sand filters, plastic sheeting (mulch), and metal frames for plastic houses and tunnels, have emerged and now occupy a place in the regional market as well.

It is important to observe the free market rules as the country deregulates prices. Laws against monopolies have been enacted and are to be strictly applied. Other laws relating to the markets have to be reviewed periodically and enforced for the protection of the consumer and small shareholders but not at the expense of investment. Competition should be encouraged to ensure that the prices are set through market forces. Credit facilities should be established and the existing one promoted to enable small farmers to purchase advanced irrigation systems and maintain them. Field pilot projects on automation should be set up or encouraged, to attract the attention of farmers and encourage their adoption. The introduction of moisture sensors to alert farmers to the need for water application and the rate thereof should be encouraged through the government's extension service (extended to farmers by the Ministry of Agriculture).

An equally important and parallel measure would be the organization of agricultural marketing and the promotion of exports, especially of Jordan Valley produce. Other crops can be introduced to the irrigated agriculture in Jordan when the markets are developed or when outside markets have a demand for such crops. In a free market environment, it is important to reward the producer by assuring him of a fair share of the agricultural income through market efficiency and organization. Lessons have been learned from past attempts and new producers should benefit from past experiences.[2]

Conclusion and policy implications

Irrigation technology was transferred to Jordan in the 1970s concurrently with government efforts to develop the Jordan Valley through an integrated social and economic plan. Water shortages resulting from the decline of the Jordanian share in the Yarmouk River and the additional need for water to expand the irrigated areas in the Jordan Valley were important factors that induced the adoption of advanced irrigation systems.

A set of other benefits facilitated the spread of irrigation technology in the Jordan Valley and on the plateau including increased agricultural yield, enhancement of treated wastewater reuse, improvement of farm economics and positive social impacts.

The water stress prompted the Jordan Valley Authority to adopt a water conveyance and distribution network capable of increasing the efficiency of its water service. Pressure pipe networks were installed in the new areas and then old areas, once served by surface concrete lined canals, were converted to pressure pipe networks. Such measures encouraged the users to adopt advanced on-farm irrigation techniques and the overall result was to mitigate the negative impact of water shortages. The government's credit arm, the Agricultural Credit Corporation, advanced long-term credit to help farmers install advanced irrigation systems on their farms. This had the advantage of water saving, improved fruit quality and reduced the need for agricultural drainage.

Contemporary pressures imposed on irrigated agriculture need government attention. They are principally generated by the removal of Treasury subsidies from fuel, increases in the cost of imported farm inputs, availability of agricultural workers, sustainability of irrigation water, agricultural marketing, and competition in local and regional markets. Farmers cannot be left alone to face these formidable challenges and the government has to devise policies and mechanisms to improve the lot of farmers.

This should have implications on the government's policy in responding to the trespassing by any co-riparian party on Jordan's water shares. It should have implications on agricultural research and extension services to help farmers adjust to the new conditions, and in maintaining partnerships in the agricultural credit institutions.

Notes

1   World countries are categorized by the World Bank according to the per capita share of the gross national income (GNI). These categories are:
(http://worldbank.org/data/countryclass/classgroups.htm):
    • low income: where the average share of the GNI per capita does not exceed $755;
    • lower middle income: where the average per capita share lies between $756 and $2995;
    • upper middle: with an average per capita share of $2996 and $9265;
    • high income: with an average share equal to $9266 or more.
    Water consumption is proportional to the standard of living which, in turn, is reflected by the per capita income.
2   The Agricultural Marketing Organization was set up under the auspices of the Minister of Agriculture in 1968 and the Jordan Valley Farmers Association, established by a separate law in 1974, was entrusted, among other responsibilities, with marketing of the Jordan Valley produce.

References

Al Rai (2008) Newspaper, 13 March, Amman, Jordan

Haddadin, M. J. (2001) *Diplomacy on the Jordan: International Conflict and Negotiated Resolution*, Kluwer Academic Publishers, Norwell, MA, pp45–46

Haddadin, M. J. (2006) *Water Resources in Jordan: Emerging Policies for Development, the Environment and Conflict Resolution*, RFF Publishers, Washington, DC, pp14, 20–21

Haddadin, M. J. (2007) 'Quantification and significance of shadow water in semi arid countries', *Water Policy*, vol 9, pp439–456

Jordan Valley Authority (1987) *The Jordan-Syria Treaty for the Utilization of the Yarmouk River*, A treaty replacing the 1953 Treaty for the same purpose, Jordan Valley Authority, Amman, Jordan

MWI (2002) *Regulation of Groundwater*, A bylaw regulating the exploitation of groundwater resources, Ministry of Water and Irrigation, Amman, Jordan

Salman, A., Al Karablieh, E., Regner, H.-J., Wolff, H.-P. and Haddadin, M. (2008) 'Participatory irrigation water management in the Jordan Valley', *Water Policy*, vol 10, no 4, pp305–322

Shibley, S. C., Bitoun, M., Gaiser, D., Nassif, H. and Schoen, V. (1987) *The Jordan Valley Dynamic Transformation 1973–1986*, Study prepared by Tech International, Inc., in association with Louis Berger International, Inc., under AID contract no ANE-0260-C-00-7054-00

9

Three Essential Elements of On-farm Irrigation Efficiency and Conservation

Baryohay Davidoff

Background

There is growing pressure on agriculture for enhanced efficiency and productivity. This pressure stems from growing competition among agricultural, urban and environmental water users for more efficient use of increasingly scarce water resources. As the world population is increasing, so is the need for additional food and fibre. Moreover, the growing urban sector of society requires ever-increasing water supplies, fuelling ever-increasing conflict and competition for water. The world population is projected to cross the 7 billion mark by 2013, 8 billion by 2028, 9 billion by 2054 and 10 billion by 2183 (United Nations, 1999). The problem is multiplied when coupled with reduced water for agriculture, increased demand for food and fibre, and enhanced life-expectancy as the standard of living improves worldwide, while expecting higher production levels per unit area of land or unit volume of water.

Like many parts of the world, California faces many water-related challenges for decades to come. These challenges directly and significantly affect, often in more than one way, how the agricultural community must use its diminishing share of water more efficiently and beneficially.

Population increase

California's population, similar to other parts of the world, is growing rapidly and putting stress on land, water, the environment, and fish and wildlife habitats. When the state's water systems were developed half a century ago with the construction of the Central Valley Project (built by the United States Federal Government, beginning in 1937) and the construction of the State Water Project in the mid-1960s, California's population was less than 20 million. In 1990 and 2000, California's population was 30 million and 36.5 million, respectively, projected to be 42.5 million and 48.9 million by 2010 and 2020, respectively (Department of Water Resources, 1994).

Environmental water uses

California will have to manage and use its scarce water resources to meet environmental, agricultural, urban and recreational needs in a sustainable manner. Protection of water quality and enhancement of the environment are important in water management and water development planning and implementation processes. When water is removed from its natural environment for urban and agricultural uses, the environment is often adversely affected. Environmental stewardship is an integral part of water management. California's diverse natural environment requires multiple stewardships to assure reliable water deliveries and maintain the health and sustainability of the environment. A major challenge for California is to maintain, restore and enhance its unique Sacramento–San Francisco Bay–Delta ecosystem. South of the Bay–Delta, 20 million people of the state, rely on the Bay–Delta system for their drinking water, and a significant acreage of agricultural land relies on the same system for irrigation water. Efforts to sustain the health of the Bay–Delta often affects water deliveries by reducing diversions, partly because of reduced precipitation and water shortages, hydrologic droughts, lack of adequate storage capacity and partly because of mandated and regulatory water shortages to protect endangered fish and wildlife species, etc.

Urbanization

Analysis of population growth in California indicates that almost the entire increase is in the urban sector. There is not only substantial residential development in existing cities and towns through infill growth, but also expansion in the number of new towns and cities. This is most visible in the Sacramento–San Joaquin Valley stretching from the most southern part of the valley to Sacramento and continuing north. The current trend of growth indicates that by the year 2030 there will be continuous urbanization stretching for more than 450km along Freeway 99 connecting the present cities of Bakersfield in the south, Fresno, Modesto, Stockton and Sacramento in the centre, to cities beyond in the north. It is estimated that California's population is growing at about 600,000 per year. Population growth and urbanization often put pressure on agricultural lands to be reclassified for urban development. The use of water associated with such lands also changes from agriculture to urban. Yet, the additional population will need additional food and fibre which must be grown with additional water.

Shifting agricultural, urban and environmental water uses

The dynamic growth and changing nature of California's population is most pronounced in the urban sector. Agriculture has been losing about 6000 hectares per year to urban and commercial development for decades. Socio-economic and environmental factors favour significant consideration for the additional allocation of water for ecosystem restoration and environmental enhancement. Analysis performed by the Department of Water Resources (2005) in the *California Water Plan Update* provides three plausible, yet very different baseline water demand scenarios by 2030. The three scenarios are water demand based on current trends, water demand based on a less resource-intensive economy and water demand based on a more resource-intensive economy. Although local and regional water management strategies and implementation of efficiency measures both in the agricultural and urban sectors will help to meet some future water demand, in order to eliminate about 1 million acre-feet (AF) overdraft of groundwater, the state may require an additional 2.5 million m^3 of water per year.

The *California Water Plan Update* (2005) analysis indicates that, under current trends of water use and conservation, while overall water use will remain about the same between now and 2030, urban water use is expected to increase by 3 million AF per year while agricultural water use will decrease by about 3.5 million AF per year, with an increase of about 250,000 AF per year for the environment. In addition, 1–2 million AF of water per year is needed to eliminate groundwater overdraft. This shift has two important implications for California. Urban water use is generally more energy intensive, therefore such transition in water use will increase CO_2 emissions and contribute significantly to greenhouse gases. In addition, to meet agricultural production pressures, i.e., agronomic, market, economic, societal and water policy pressures and stresses, the agricultural sector must become more efficient and productive.

California's 9 million acres of agricultural lands rely on both surface- and groundwater for irrigation. More than 200 irrigation districts manage and distribute about 35 million AF of water annually (surface- and groundwater) to growers through thousands of local reservoirs and regulating reservoirs, and thousands of kilometres of open channels and pipelines. Agricultural lands are irrigated by means of: furrow; flood, basin or border irrigation; solid set and movable sprinkler systems; buried and surface drip systems; micro sprinklers; centre pivot, etc. Excluding rice, more than 60 per cent of irrigated land in California is under furrow irrigation systems (Department of Water Resources, 2005).

California's response to increased water scarcity

Water is the lifeline of California's growth, economy and prosperity. California's agriculture, environment, cities, industries (including high-tech computer and entertainment) and tourism industry are all heavily dependent on an adequate and reliable supply of water. California has been in the forefront of the development of complex and sophisticated water storage, distribution and delivery systems. Likewise, overall, the sophisticated and complex systems of water use in California are a result of: growers' hard work; timely availability of services, transportation, functional markets, services of private industry; and technical and informational support by the state's educational systems including extension services, farm advisers and irrigation specialists.

California has the statutory requirement that all water right holders must use water *reasonably and beneficially*:

> *Water resources of the State be put to beneficial use to the fullest extent of which they are capable, and that the waste or unreasonable use or unreasonable method of use of water be prevented, and that the conservation of such water is to be exercised with a view to the reasonable and beneficial use thereof in the interest of the people and for the public welfare.* (California Water Code)

Wasteful and non-beneficial use of water can result in loss of water rights. As a result, California's diverse agricultural sector has one of the most advanced and modern water use and delivery systems in the world. Indeed, California's efficiency in the productivity of food and fibre is demonstrated by the state's relatively high irrigation efficiency and high yields per acre of land compared with the rest of the world.

Specifically, beginning with the droughts of the late 1970s, prolonged droughts and water shortages of the early 1990s and current water shortages due to either hydrologic droughts or legislative water shortages, California has made incredible strides forward in the efficient and beneficial use of water resources. These efforts, in addition to legislative and statutory requirements, have included a massive investment of public funds for technical and financial programmes to further advance water use efficiency.

Hoagland and Davidoff (1998) fully explored agricultural water use policy issues in California. The intent of this chapter is to further discuss policies and practices that help improve the three essential elements of prudent on-farm efficiency and conservation. The Californian state government and the agriculture industry have developed a unique approach to further advance water use efficiency in the state. These efforts have been implemented at all levels, i.e., state, regional and irrigation districts, which operate, maintain and manage water systems, and distribute and deliver water to individual growers who manage water application at the on-farm level.

Legislative requirements

A formal agricultural water management planning process began in 1989 with Assembly Bill (AB) 1658, which required all irrigation/water districts distributing 50,000 AF of water or more per year to prepare a Water Management Plan (WMP) and identify any opportunity to reduce selenium-contaminated drainage water flows or any water conservation opportunities. In subsequent legislation in 1991, AB 3616 required that the state develop a list of agricultural Efficient Water Management Practices (EWMPs) in cooperation with irrigation/water districts and assist water suppliers in developing WMPs.

The WMP identifies all cost-effective EWMPs that must be implemented and develops a schedule for implementation. Water suppliers may apply for exemption from having to implement a particular EWMP, but only based on a comprehensive net benefit analysis. Net benefit analysis is a comprehensive process that includes engineering benefit to cost analysis as well as any quantitative and qualitative analysis of the environmental, third party, social, economic and financial impacts of implementing any given EWMP. The work of AB 3616 has evolved into the formation of a non-profit Agricultural Water Management Council (www.agwatercouncil.org) which oversees the development of agricultural WMPs and the implementation of EWMPs. The state has continued to participate in the AB 3616 process by providing financial and technical assistance and by providing an analysis of all WMPs submitted to the council. More than 79 major water suppliers and irrigation districts, constituting about half of the state's irrigated agriculture, are signatories to a *Memorandum of Understanding Regarding Efficient Agricultural Water Management Practices for California Water Suppliers*. The signatories to the AB 3616 process and Memorandum make a voluntary commitment to develop WMPs and implement all EWMPs (Table 9.1) that are cost effective based on the net benefit analysis (AWMC, 1999).

Financial assistance

As a matter of public policy, California has developed a comprehensive programmatic, technical, financial and policy framework for addressing water management challenges. The *California Water Plan Update* (2005) provided recommendations for actions to be taken during the next 25 years. Among these recommendations are that California must invest in reliable, high quality, sustainable and affordable water conservation, efficient water management, flood control, and development of water supplies to protect public health and improve California's economy, environment and standard of living. The cost to implement agricultural water use efficiency recommendations alone is estimated to range from US$0.3 to US$4 billion over the next 25 years. These funds will be raised mainly through voter approved Water Bond measures, and will primarily be grant funds for projects that will provide water efficiency benefits to the state. For the past five years, more than US$80 million has been committed by the state to help

Table 9.1 *List of EWMPs developed under AB 3616 for California agricultural water suppliers*

1	Prepare and adopt a WMP
2	Designate a water conservation coordinator
3	Support the availability of water management services to water users
4	Improve communication and cooperation among water suppliers, water users, and others
5	Evaluate the need for changes in policies of the institutions to which the water supplier is subject
6	Evaluate and improve efficiencies of water suppliers' pumps
7	Facilitate alternative land use
8	Facilitate use of available recycled water
9	Facilitate the financing of capital improvements for on-farm irrigation systems
10	Facilitate voluntary water transfers
11	Line or pipe ditches and canals
12	Increase flexibility in water ordering and delivery
13	Construct and operate water supplier spill and tail-water recovery systems
14	Optimize conjunctive use of surface- and groundwater
15	Automate canal structures
16	Water measurement and water use report
17	Pricing or other incentives

advance agricultural water use efficiency programmes. The Department of Water Resources (DWR) has developed a comprehensive competitive Water Use Efficiency Grant Program to solicit project proposals for funding.[1]

Essential elements of on-farm efficiency

The State of California provides technical and financial assistance to advance the three essential elements of on-farm irrigation efficiency improvements and water conservation: providing irrigation scheduling information, improving irrigation water distribution and improving water availability on an on-demand basis. The assistance, specific to efficiency improvements at both water supplier and on-farm levels, comes through three important technical and financial assistance programmes:

1   Crop water use information: The California Irrigation Management Information System (CIMIS) provides reference evapotranspiration and crop coefficient information to estimate crop water use which is essential for irrigation scheduling.

2   Improved distribution uniformity (DU): The On-Farm Irrigation System Evaluation Mobile Laboratory provides for irrigation water distribution uniformity improvements. DU determines irrigation applied water requirement.

3   On-demand water delivery systems: The Water Use Efficiency Grant Program provides for on-demand and flexible water delivery system improvements through public grant funds.

The CIMIS programme is a basic programme that is budgeted, funded and administered by the Department of Water Resources. The On-Farm Irrigation System Evaluation Mobile Laboratory programme and the on-demand water delivery system improvement activities are funded through state grants and loans to local irrigation districts and other local agencies.

Crop water use information: CIMIS

The first essential element for efficient irrigation is the timely availability of quality crop water use data. Irrigators must have information to determine when and how much water to irrigate. Many methods and approaches are available to growers including estimation of crop water use based on pan evaporation, air temperature, lysimeters, the Blaney–Cradle method, the Penman–Monteith method, and a few other methods.

The State of California has developed CIMIS – a comprehensive data and information collection, analysis and dissemination system based on climatological factors. CIMIS is a network of more than 130 standardized, computerized and automated weather stations located throughout the state, which are connected to and communicate with a central computer in the DWR in Sacramento. The weather stations collect minute-by-minute climatological data such as air temperature, relative humidity, solar radiation, wind speed and direction. Figure 9.1 shows a typical CIMIS station with all instruments.

Following a comprehensive data quality control, the central computer calculates the reference evapotranspiration (ETo) using the Penman–Monteith energy

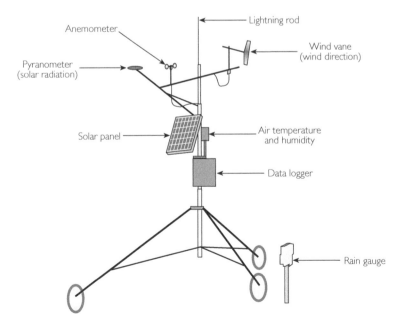

Figure 9.1 *A typical CIMIS weather station with standard instrumentation*

balance method. The technical details of the procedures to collect and calculate ETo are documented in *Technical Elements of CIMIS* (Department of Water Resources, 1998). ETo is defined as evaporation from soil and transpiration from well-watered grass, as a reference base. When the central computer is called, data for all reporting stations are available free of charge the following day, for all prior days as long as the station has been in operation.

ETo is the amount of water that has been lost from a field that must be put back by irrigation. For any given location, crop and time period, crop water use could be higher or lower than ETo. An adjustment must be made to ETo using crop coefficients (Kc). Kc values have been developed for many trees, vines, agronomic crops, vegetables and landscape plants. ETo and Kc values vary during the growing season from planting time to harvest. With these values, crop water use, ETc, can be calculated and the amount of water that must be put back into the root zone can be estimated for any particular crop. Equation 1 provides the first estimation of crop water use based on ETo and Kc:

Crop water use = ETc = ETo × Kc   (Equation 1)

In addition to water needed for crop use, ETc, growers must take into consideration other factors such as the leaching requirement (LR) to maintain a sustainable soil environment for optimum crop growth, and water that may be needed for frost or weed control, etc., as additional beneficial uses of water. The LR is the amount of water in addition to crop water use that is needed to prevent the accumulation of salt and to maintain a sustainable salt balance in the root zone. Depending on the salinity level in the soil root zone and irrigation water quality, the LR may range from 5 to 15 per cent of crop water use (Davidoff et al, 1996).

In this chapter, beneficial use of water will be limited to evapotranspiration (ET) and LR. Therefore, crop water requirements can be estimated by Equation 2.

Benefical water use = ETo × Kc + LR   (Equation 2)

Irrigation efficiency (IE) is defined in Equation 3.

$$\text{Irrigation efficiency} = \frac{\text{Beneficial water uses}}{\text{Applied water}} \times 100 \qquad \text{(Equation 3)}$$

Applied water is the gross amount of water that is actually put on a field with a given irrigation event. Applied water is always impacted by how uniformly irrigation systems distribute water over a field. Irrigation distribution uniformity is the second essential element of on-farm irrigation efficiency. It is essential to understand IE in the context of on-farm and regional levels. Solomon and Davidoff

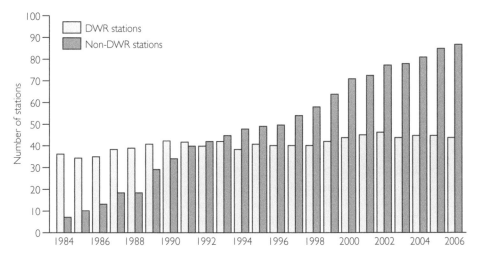

Figure 9.2 *Growth of the CIMIS weather stations*

(1999) showed that unlike on-farm IE, regional IE almost always includes reuse of water (surface run-off and deep percolation) from on-farm irrigation, thus achieving higher regional efficiencies through water reuse.

CIMIS funding

Since 1985, the number of CIMIS stations, registered users, and calls and enquiries to the system have more than quadrupled. Figure 9.2 shows the growth of CIMIS stations. California's DWR established and owns about 40 stations; the remaining 90 stations are funded, installed and maintained by local agencies. The DWR annually calibrates the instrumentation for all stations and provides data quality control. The DWR also maintains the CIMIS database and makes it available on its website. The programme is a true partnership between the state and local agencies.

CIMIS users

CIMIS data are accessed by various users, including growers and irrigators; irrigation consultants; irrigation and water districts; public agencies; universities and other state educational institutions; WMP developers; urban water users; landscape, parks and large turf area water users; and others (fire fighters, National Weather Service, litigators, State Integrated Pest Management, etc.).

CIMIS ETo information dissemination

CIMIS daily, weekly and monthly ETo information is available directly on the DWR's website, www.cimis.water.ca.gov. CIMIS users can access the CIMIS computer directly to obtain data and information. Many organizations obtain the CIMIS ETo data and distribute them to wider users, including: Center for Irrigation Technology at California State University Fresno; agricultural and urban irrigation/water districts; resource conservation districts; landscape water users; radio and television stations; telephone recordings; newspapers; and consultants.

For the year 2000, it was estimated that there were more than 80,000 calls and enquiries made directly to the CIMIS computer. Since the advent of the internet and availability of the ETo data on the CIMIS website and other dissemination points, there has been an explosion in the use of CIMIS and ETo data.

CIMIS benefits

Parker et al (2000) conducted a two-tiered survey and study on CIMIS benefits. The survey covered about 363,816 acres of irrigated agricultural lands, covering many different regions (22 counties) and crops. The survey also included residential irrigators who use large amounts of relatively expensive treated water. Residential landscape, citrus and avocado growers had the largest per-acre benefits among CIMIS users. These water users along with golf courses, parks and cemeteries could take the most advantage of CIMIS ETo data since they usually use sophisticated irrigation systems and expensive water. Analysis of the CIMIS users' survey data showed that this publicly funded programme, with the state's expenditure of about US$800,000 per year, provided about US$64.7 million in benefits annually state-wide. The results of the survey showed that agricultural water use was reduced by 107,300 AF annually. The growth and benefits of the CIMIS programme are summarized in a DWR (1997) publication.

In addition, agricultural and urban water agencies are increasingly using CIMIS ETo data to develop water balance and water budget information. These water balances and water budgets are required as an integral part of Urban and Agricultural Water Management Plans.

Distribution uniformity: On-Farm Irrigation System Evaluation Mobile Laboratory

The second essential element of prudent on-farm water management is to improve the DU of irrigation systems. No irrigation system is capable of applying water at 100 per cent uniformity, that is, delivering equal amounts of water to all parts of the field. The deviation, expressed as a ratio, is called distribution uniformity and is defined as:

$$DU = \frac{\text{Average low quarter depth of water applied to plants in a field}}{\text{Average amount of water applied to plants}} \times 100 \qquad \text{(Equation 4)}$$

Non-uniformity in water application and distribution is due to the inherent limitations in the manufacturing, design, operation, maintenance and manage-

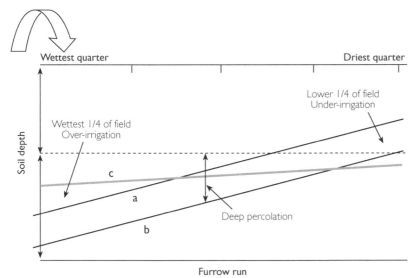

Figure 9.3 *A typical distribution uniformity in a furrow irrigation*

ment of irrigation systems. Parts of a field receive more and other parts receive less than the ideal required amount of water. For example for furrow irrigation, Figure 9.3 demonstrates three different DUs depicted with lines a, b and c. Due to differential infiltration opportunity time, portions of a typical furrow closer to the water source will systematically be wetter to depths much deeper than the root zone and over-irrigated, while parts of the furrow more distant from the water source will systematically be drier and under-irrigated.

Irrigators often try to make sure that no parts of a field are under-irrigated (line a) and plants in the driest part of the field receive adequate irrigation water to meet the crop water requirement. This management decision almost always causes some parts of a field, already over-irrigated, to receive even more irrigation water resulting in a gross inefficiency in water use and excessive drainage (line b). Improving DU will reduce excess deep percolation, as depicted by line c. Study and analysis by Sanden et al (2003) demonstrated the impact of DU on the depth of applied water received at the wettest, wet, drier and dry quarters of a field and on the yield of alfalfa under different total average applied water. For example, as shown in Table 9.2, with a DU of 70 per cent and 42 inches of applied water, the wettest, wet, drier and dry portions of the field received 55, 46, 38 and 29 inches of water, respectively. The analysis shows that for an average 42 inches of applied water, half of the field received between 38 and 46 inches of water, the wettest quarter received 13 inches more water than needed and the dry quarter received 13 inches less water than needed. The yields corresponded accordingly to the levels of applied water.

In this example, if an irrigator decides to make sure that the dry quarter received 42 inches of water to satisfy the crop water requirement, without improving the DU, the wettest quarter will receive in excess of 36 inches of

Table 9.2 *Estimated effective depth of applied water over different quarters of an alfalfa field with 70, 80 and 90 per cent distribution uniformity (DU) and the resulting hay yield using the 'average' San Joaquin Valley production function*

		Qtr. irrig. by avg. depth (in.)				Qtr. yield by avg. depth (t/ac)			
70%	Field qtr	42	48	54	60	42	48	54	60
DU	Wettest	55	62	70	78	8.5	7.6	6.0	5.0
	Wet	46	53	59	66	8.2	8.6	8.1	6.7
	Drier	38	43	49	54	6.6	7.8	8.5	8.5
	Dry	29	34	38	42	3.6	5.3	6.6	7.6
	Field average yield (t/ac:):					6.7	7.3	7.3	7.0
80%		42	48	54	60	42	48	54	60
DU	Wettest	50	58	65	72	8.5	8.3	7.0	5.9
	Wet	45	51	58	64	8.1	8.6	8.3	7.2
	Drier	39	45	50	56	7.0	8.1	8.5	8.4
	Dry	34	38	43	48	5.3	6.8	7.8	8.4
	Field average yield (t/ac):					7.2	7.9	7.9	7.5
		42	48	54	60	42	48	54	60
90%	Wettest	46	53	59	66	8.2	8.6	8.1	6.7
DU	Wet	43	50	56	62	7.8	8.5	8.4	7.6
	Drier	41	46	52	58	7.3	8.3	8.6	8.2
	Dry	38	43	49	54	6.6	7.8	8.5	8.5
	Field average yield (t/ac):					7.5	8.3	8.4	7.8

Source: Sanden et al (2003)

water, a gross inefficiency. By improving the DU to 90 per cent, the application of 42 inches of water will result in over-irrigation of the wettest quarter by 4 inches and under-irrigation of the dry quarter by 4 inches. Improving the DU is essential for optimal on-farm irrigation management.

DU has a significant impact on the gross amount of water that needs to be put on a field. Likewise, IE is also directly impacted by DU; the higher the DU, the higher the IE. The DU and IE relationship assumes that applied water provides adequate water within the root zone without generating excess deep percolation.

To determine the gross amount of water that must be applied to a field to meet crop water consumption, the irrigator must consider the DU of an irrigation system as follows:

$$\text{Applied water} = \frac{\text{ETo} \times \text{Kc} + \text{LR}}{\text{DU}} \qquad \text{(Equation 5)}$$

ETo (and crop coefficient, Kc) provides fundamental information for irrigation scheduling, i.e., when to irrigate and how much water to apply, which is necessary for on-farm irrigation efficiency. However, the DU controls and determines the gross amount of water that must be applied. If Eto × Kc + LR is 30 inches

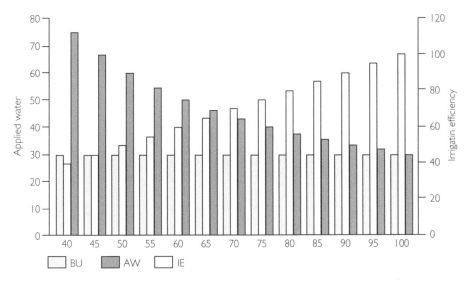

Figure 9.4 *Relationship between distribution uniformity, irrigation efficiency, and applied water, based on 30 inches of beneficial water use*

for a crop through the growing season, the applied water must be 42.8 inches for an irrigation system that has a DU of 70 per cent. DU determines IE, in this example 70 per cent. Gross applied water will be 37.5 inches by increasing DU to 80 per cent, a reduction of gross applied water by 5.2 inches, or a 12.3 per cent conservation. Water conservation and savings will be 10.8 inches by improving DU for hand-move sprinklers from 62 to 80 per cent. Figure 9.4 demonstrates the relationships between DU, applied water and IE. This analysis is for a crop that requires 30 inches of water as total beneficial water use (BU).

Sources of DU

Different factors may cause a low DU in different irrigation systems. For example, even a well-designed sprinkler irrigation system inherently has some non-uniformity. There is variation in water flow from different sprinkler heads or nozzles. This variation in volume of water discharged from each sprinkler nozzle could be due to variations in the hydraulic characteristics of nozzles, pressure differentiation and head losses along the water delivery main line and laterals. Also, variation in spacing between nozzles along an aluminium pipe, variation in spacing between parallel aluminium laterals, tilt of sprinkler risers (from perpendicular to the laterals) and variation in water flow with time, are common. Often variations in flow are a result of mixed nozzle types. For furrow irrigation, non-uniformity is systematic as the start of furrows receives more water than the end, due to the variation in water infiltration opportunity time between ends of the furrow run. For example, to achieve a required 10 hours of irrigation time at the end of a furrow, water may have to run for 14 hours at the beginning of the furrow. In addition, some variability may be introduced due to

Table 9.3 *Distribution uniformity for several irrigation systems**

Irrigation method	Number	Mean (%)	SD (%)	CV(%)
Continuous-move Sprinklers	57	75 a**	10	13
Hand-move sprinklers	164	62 c	15	24
Under-tree sprinklers	28	79 ab	16	20
Furrow	157	81 b	15	19
Border	72	84 b	14	17
Micro-irrigation	458	73 a	15	21

Note: * Unpublished DWR data from irrigation system evaluation programme reports
** No significant differences in mean values designated with same letters

the spatial and temporal variability of soil characteristics, including topography, water infiltration rates, soil texture and soil structure, etc.

An analysis of more than 900 field evaluations of distribution uniformity for a variety of irrigation systems and different regions of California is given in Table 9.3.

Advanced technology and pressurized irrigation systems, such as drip and sprinklers, provide the greatest control over the amount of water applied, and thus inherently have higher theoretical distribution uniformity. Experience in California has shown that irrigators rely on the hardware and software components of these irrigation systems, but in practice are often deficient in good system management. On the other hand, furrow and border irrigation systems, which theoretically have a lower DU, receive a higher level of management. Irrigators can visually inspect the water flow and advance of water in furrow and border irrigation systems, and make timely and appropriate management decisions, such as land levelling and preparation, changes in flow rates, etc. Given that hands-on observation and adjustment is more limited with drip and sprinkler irrigation systems, this is stipulated as one explanation as to why gravity-fed irrigation systems have relatively higher empirical DUs compared with pressurized irrigation systems. In addition to design criteria, operation, maintenance and skilful management are of paramount importance for more sophisticated, high technology, pressurized irrigation systems to achieve their theoretically higher DU capability. Therefore, improvements in system DU provide the greatest opportunity for increased efficiency and conservation.

Pitts et al (1996) reported results from 385 irrigation system evaluations. Their analysis indicated that more than 80 per cent of the evaluations resulted in recommendations for improvements, and in over 90 per cent of evaluations growers implemented at least one of the recommendations for improvements in DU. They showed that mean DU varied with irrigation system type with 65, 74, 65, 70 and 49 per cent for agricultural sprinklers, drip, micro-sprayers, furrow, and non-agricultural (turf) sprinklers, respectively. A follow-up irrigation system evaluation after recommendations were implemented by growers showed an average improvement of 18 per cent in DU.

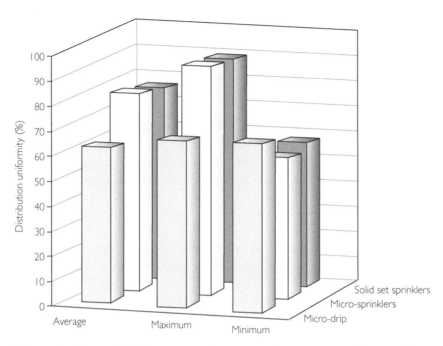

Figure 9.5 *Range of DU variation and opportunity for improvements for micro-drips, micro-sprinklers and solid set sprinklers (data from 102 on-farm irrigation system evaluations over 8051 acres of trees and vines)*

Distribution uniformity data from another mobile laboratory reported by Hockett (2006) comprised the results of distribution uniformity (DU) evaluations conducted over 8051 acres of irrigated fields. These DU evaluations were mainly carried out on trees and vines on 47 micro-drips, 50 micro-sprinklers and four carrot fields under solid-set sprinkler irrigation systems. These results are shown in Figure 9.5. Again, range of DU variation shows potential for improvements and fine tuning of these systems to increase the DU and improve on-farm irrigation efficiency.

On-farm irrigation system evaluation mobile laboratory

The mobile laboratory is a unique technical assistance programme that utilizes field irrigation data such as flow rates, system characteristics, pressures, soil properties, specific computer software and hardware, and management decision-making techniques and processes to analyse the causes and sources of irrigation non-uniformity and provide simple and quick recommendations to improve hardware and management practices. The goal is to facilitate irrigation system management for optimal performance and distribution of water over the field as uniformly as design specifications will allow.

The mobile laboratory consists of two or three technical staff (a team leader and one or two assistants), a vehicle loaded with the necessary equipment to take field measurements such as soil moisture, flow rate, volume and pressure measurement, and a computer with specific irrigation software. The mobile laboratory team leader contacts irrigators, and schedules an on-farm irrigation system evaluation during the growing season. Often, growers contact the mobile laboratory and request assistance. Mobile laboratory services are free of charge to the growers.

The mobile laboratory is run as a partnership between the state, resource conservation districts, local irrigation districts and irrigators. The Irrigation Training and Research Center (ITRC) (www.itrc.org) at California Polytechnic State University in San Luis Obispo, California, played a crucial role in the development of mobile laboratory procedures and training staff. The goals of the mobile laboratory partnership are to:

- perform on-farm irrigation system evaluations during irrigation events;
- collect irrigation system-specific information and data pertinent to single irrigation events;
- analyse the information and data to determine DU;
- analyse the information and data to determine causes and sources of non-uniformity in water distribution systems;
- make easy-to-implement recommendations to growers for changes in the management or hardware to improve DU; and
- for selected fields, perform a follow-up irrigation system evaluation after recommendations are implemented by the irrigators.

The mobile laboratory programme was started in the early 1980s with funds from the California DWR. Each mobile laboratory costs about US$150,000 annually with a three-year cooperative agreement. The success of the programme and high interest and participation by irrigators has necessitated its continuation. This has been possible through cost-sharing by local irrigation districts. The DWR funded the programme at 100, 80, 50 and 20 per cent levels as the local irrigation district cost-share level rose through 0, 20, 50 and 80 per cent levels for the initial three-year and three consecutive three-year periods, respectively. Currently, eight mobile laboratories are fully funded by local irrigation districts throughout the state. Depending on the size of the districts and number of irrigators, between 5 and 15 local irrigation districts participate in supporting and funding a single mobile laboratory. Depending on the number of staff, each mobile laboratory can conduct between 50 and 80 evaluations per year. Local resource conservation districts (as part of the United States Department of Agriculture) initiate and run the programmes, including solicitation of funds from irrigation districts. For confidentiality purposes, while data and information from each individual field and grower are shared with that grower, only aggregate information is available to the public.

A typical mobile laboratory consists of:

- a team leader;
- one or two assistants;
- a vehicle (truck);
- soil, water and plant measurement tools;
- computer and software;
- resource conservation district operation;
- participation of between 5 and 15 irrigation districts.

A typical irrigation system evaluation includes the following:

- measuring all possible variables affecting the amount of water received by plants across a particular field;
- computing the DU of a single irrigation event;
- computing the potential seasonal irrigation efficiency based on DU; and
- providing a one-page printout to the irrigator summarizing possible water and dollar savings.

Some typical irrigation system evaluation recommendations made to the growers include:

- replace worn sprinkler nozzles and nozzles mixed with different brands;
- use pressure regulators where needed;
- use sand media filters in drip and sprinkler systems;
- prevent plugging of drip emitters and sprinkler nozzles, and flush pipes;
- keep plant roots from intruding into drip and sprinkler pipes;
- prevent large pressure variations in pressurized irrigation systems;
- minimize variation in furrow infiltration opportunity time;
- optimize flow and water advance rates in surface irrigation systems, including surge systems;
- modify irrigation timing and application rates;
- develop and manage run-off recovery systems for one or more fields;
- combine surge irrigation with furrow systems;
- use the expertise of private consultants who can evaluate and help improve irrigation systems;
- change irrigation systems when current systems are not appropriate for the crop, soil and topography.

Irrigation system evaluation training and education workshops
Since the mid-1980s, the importance of irrigation system improvements to achieve higher DUs has been recognized, as well as the value of a state-developed comprehensive training and education programme. The purpose of the training and education programme is to:

- develop a uniform and consistent, yet simple, functional and farmer-friendly programme, along with materials, procedures and software for the evaluation of each irrigation system;
- conduct workshops to train mobile laboratory staff and water/irrigation district staff and others on how to perform irrigation system evaluations;
- continue to provide technical support including use of software.

The ITRC, in partnership and with financial support from the DWR, has developed an educational and training workshop for irrigation system evaluation. Since 1985, the DWR has been funding this adjunct to the mobile laboratory. The ITRC conducts 2, two-and-a-half-day hands-on workshops annually. These irrigation system evaluation workshops have been immensely successful with each workshop filled to capacity by attendees every year. Attendees of these workshops typically include mobile laboratory staff, irrigators and water and irrigation district staff, growers, large farming operation staff, agricultural consultants, government agency staff who want a comprehensive understanding of on-farm water management, and other interested individuals.

The attendees are trained on how to conduct irrigation system evaluations, analyse the information and data, determine DUs and provide easy-to-implement recommendations to irrigators.

On-demand water delivery systems:
Water Use Efficiency Grant Program

On-demand water delivery system is the third and key component of on-farm irrigation management and efficiency. In the absence of the availability of water when and where it is needed, any irrigation scheduling and irrigator's knowledge of crop water requirements, crop coefficients and irrigation system performance will arguably be of very limited use.

Growers/irrigators that rely on groundwater pumping can have total control over all three components of on-farm irrigation management and efficiency through real time and on-demand availability of water. Irrigation water is available upon turning on a pump. Likewise, termination of irrigation is a real time management decision by turning off the pump. The availability and delivery schedule is different for growers/irrigators that rely on water deliveries from irrigation districts or diversions from surface water resources and streams. Such water deliveries are subject to many constraints such as system capacity, infrastructure limitations, precipitation, legal issues, water rights, droughts, water shortages, environmental concerns and water allocation schedules.

In urban and residential settings in many countries, water users can turn on the tap and water will run immediately. Not so in the agricultural setting. Even in the most advanced countries that have sophisticated irrigation water storage, conveyance, distribution and delivery systems, water deliveries to irrigators are often on a rotational basis. When irrigators request water, they must wait at least

24 to 72 hours or more likely one to two weeks for delivery. In many parts of the world, water availability/delivery is often on a three-week rotational basis or even longer. Under such rotational water delivery systems, and due to uncertainty of weather conditions, growers/irrigators have to apply as much water as they can get regardless of soil moisture conditions, plant water needs, previous rainfall, etc. Growers/irrigators consider this practice as reasonable insurance against crop failure when timely availability and quantity of water is uncertain.

California's diverse hydrologic regions with diverse hydraulics of irrigation districts, necessitates diverse ways of providing irrigation water to growers/ irrigators. In some areas, combined availability and delivery of groundwater and surface water adds to the complexity of on-demand water delivery. The range of water delivery systems and schedules in California are given below:

- on demand (groundwater);
- within 24 hours, very few irrigation districts;
- within 48 hours, common schedule;
- within 72 hours, common schedule;
- within 7 days, a few irrigation districts;
- within 15 or 16 days, very few irrigation districts.

Even in California with its sophisticated and complex water systems, most water deliveries are within 48 to 72 hours from the time requested. Information from a particular irrigation district delivery system with an innovative water pricing policy shows more flexibility in water ordering and delivery, as follows:

1   on-demand delivery without notification to the irrigation district;
2   arranged on-demand delivery upon irrigator's request for water;
3   constant/continuous flow delivery to large farms with multiple fields and irrigators deciding how to rotate water delivery among fields;
4   other delivery based on irrigators placing an order, with normal delivery within four days.

The district reports that between 75 and 80 per cent of water orders fall under category 4 of water delivery, i.e., irrigators begin receiving the requested amount of irrigation water within 56 hours.

Aside from legal and legislative constraints on water allocations and schedules of water deliveries, infrastructure is one of the main limitations for on-demand water delivery. Infrastructure may include many diverse aspects of water storage, transmission and delivery systems. The incapability of most water storage, conveyance, distribution and irrigation delivery systems to deliver irrigation water on demand stems from two important factors. The first factor is that for technical reasons, no irrigation conveyance and delivery system can be large enough to make water available on demand to all growers at the same time. The second factor is that it is economically prohibitive to have water distribution systems large enough to meet large agricultural irrigation water needs for all growers on demand and simultaneously. Large amounts of water are often transferred through canals and aqueducts over long distances. It takes time for water to reach its destination in the right amount. To eliminate this problem, reservoirs with large storage capacity and many smaller regulatory reservoirs must be constructed near where water is needed. The cost of such infrastructure is often economically prohibitive.

Among many efficiency improvements, the following examples summarize improvements to increase on-demand water delivery capability in the state:

- improvements in local system capacity;
- construction and/or enlargement of local storage reservoirs;
- construction of regulatory reservoirs;
- automated flow control structures;
- lining or piping of earth ditches and canals;
- tail-water and spill-recovery systems;
- improvements to in-canal-storage capacity;
- canal automation including Supervisory Control and Data Acquisition (SCADA) systems; and
- flexible water delivery systems and canal water level fluctuation controls.

In California, it is recognized that improvements in water delivery infrastructure will make surface water deliveries as close to on-demand delivery as possible. The necessity of public fund investment is evident from the fact that many projects may benefit not only local irrigation districts, but they may have regional and state benefits as well. Thus the state has an interest in partially or fully funding water use efficiency projects. General criteria for state funding of water use efficiency projects require water savings, reduction in applied water, enhancement of flow and timing, water quality improvements and energy improvements. California voters have approved several Water Bond measures to provide loans and grant funds to local water agencies to advance water management in the state. Starting in 2001, Senate Bill 23 was followed by Proposition 13 Water Use Efficiency Grant Programs of 2001, 2002, 2003 and Proposition 50 Water Use Efficiency Grant Programs of 2004, 2007 and 2008. California has invested more than US$70 million in agricultural water use efficiency improvement measures. A total of US$130 million has been invested in more than 300 individual projects for both urban and agricultural water use efficiency improvements. Depending on the type of project, the cost of each project to the state ranges from US$100,000 to US$3 million. It is important to note that only about a quarter of the projects applying for funding have actually been funded. In addition, more than US$30 million has been made available for low interest loans to irrigation districts to improve agricultural water use efficiency. A significant amount of funds have been invested in projects that improve irrigation water management, canal automation, canal flow control, canal lining or piping and water delivery systems, both at the irrigation district and on-farm levels. Along with state funding, many irrigation districts continuously invest in appropriate improvements at enormous cost. In addition, growers spend hundreds of millions of dollars each year for improvements in on-farm water management.

Criteria for investment of public funds for water use efficiency improvements:

- water savings;
- applied water reduction;
- in-stream flow and timing;
- water quality;
- energy improvements;
- other benefits.

Projects that have water savings and one or more of the above criteria are eligible for funding. US$130 million has been invested on these items (2001–2008).

Figure 9.6 shows project types, total state share of funding, and number of projects that have been funded with public funds. All of these projects help to further advance water management and achieve on-demand water delivery systems.

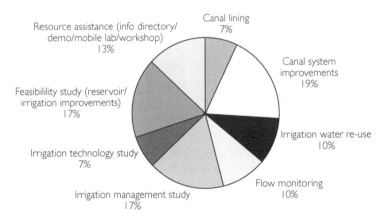

Figure 9.6 *Project types and per cent state share of funding, and number of projects funded with public funds*

Rapid Appraisal Process

Among the many projects funded under the Water Use Efficiency Grant Program is a unique and innovative project to assist the transfer of technology to irrigation districts. The project called Rapid Appraisal Process (RAP) was developed by the ITRC. The goal of RAP is to provide a systematic, innovative

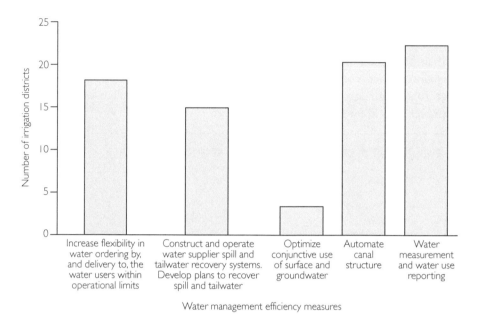

Figure 9.7 *Type and frequency of recommendations for 25 irrigation districts evaluated with RAP conducted by the ITRC*

and quick look at an irrigation district's water management issues, limitations and opportunities. Based on the RAP, the ITRC provides an irrigation district with recommendations to improve the district's water management. In a recent three-year RAP project, the ITRC evaluated 25 irrigation districts encompassing more than 1 million acres of irrigated land in California.

Figure 9.7 shows the frequency and type of recommendations made for these 25 irrigation districts. Figure 9.7 also clearly indicates that all of the districts under evaluation can be improved by implementing one or more of the efficiency measures. The majority of irrigation districts would benefit from canal automation, flexible water delivery, and better management of spills and tail-water return systems, as well as water measurement and water use reporting. The RAP project's cost to the state was about US$15,500 per district. As seen from Figure 9.7, the need for irrigation district improvements is enormous, requiring a large investment of public and private funds.

Conclusion

The pressure of ever-increasing population intensifies competition for scarce water resources among urban, agricultural and environmental water users. At the same time, the need for additional food and fibre production necessitates efficient water use. The agricultural community must become as efficient in its use of water as possible.

California experience

The California experience has shown that improved on-farm irrigation efficiency to conserve water and eliminate waste is a complex effort. Three essential elements of efficient on-farm irrigation management must be in place to enable growers/irrigators to accomplish the goal of water conservation:

1 accurate crop water requirement information for the crops being grown;
2 a fine-tuned irrigation system suitable for achieving the highest possible irrigation water distribution uniformity; and
3 an on-demand irrigation water supply capable of being delivered at the time and in the quantity needed.

Development of specific farmer-friendly technical and financial programmes is required to deliver and implement these three specific efficiency elements. The absence of any one of these three elements makes it very difficult for growers to achieve on-farm water conservation and use efficiency. In California, on-farm efficiency and conservation are achieved through the providing of services such as the CIMIS and On-Farm Irrigation System Evaluation Mobile Laboratories, along with financial and technical assistance through commitments of significant public funds as a matter of public policy oriented towards better stewardship of water resources. The uniqueness of these programmes is in their use of advanced technology, principles and practices of irrigation management, outreach, and cooperative and collaborative efforts.

Potential application in other countries

The principles and practices of irrigation efficiency are the same for irrigated agriculture all over the world. Yet water management issues are more complex in developing countries that may have limited financial and technical resources, where water storage, transfer and distribution systems may not be conducive to improved efficiency practices. Under such circumstances, growers face even greater challenges with uncertainty over water deliveries for their next irrigation. Therefore, they simply cannot plan irrigation scheduling and efficiency, and conservation is a major challenge. Growers cannot reasonably be expected to be efficient if they do not have accurate and timely information, operate poorly performing irrigation systems, or do not have water when needed.

Just as California has done over the past three decades, other countries must address the three essential elements of irrigation efficiency discussed in this chapter. Other countries can benefit from the California experience in developing, financing, delivering and administering similar essential efficiency programmes. Other countries may use California's CIMIS, mobile laboratory and infusion of significant amounts of public funds as examples to develop similar programmes to achieve these three critical and essential on-farm efficiency elements at a scale that will fit the need of the given agricultural setting. They may even surpass the Californian experience by incorporating newer technologies such as remote sensing, global positioning systems and precision irrigation technologies in on-farm irrigation management programmes. This will in part depend on the level of financial commitment and investment of public funds.

Policy implications

On-farm efficiency and conservation involve not only technical aspects of irrigation principles and practices, but also involve commitment of public funds and clear public policies, mandates and the establishment of roles and responsibilities for resource stewardship and water management. Development and implementation of on-farm efficiency elements involves overlapping roles and responsibilities of growers, water suppliers, government agencies and others. Public and private functional partnerships must be developed among growers, water districts, private industry, educational institutions, public interest groups and other stakeholders, as a matter of public policy. Specifically, these partnerships must foster cooperation between public policy makers, water managers, growers, private agricultural and irrigation industries, financial institutions, educational systems, farm advisers and irrigation specialists. Only through these functional partnerships can the complex issues related to on-farm water management, efficiency and conservation be addressed.

Clear mandates and expectations for stewardship of scarce water resources along with significant investment of public funds in water management as a matter of public policy are needed more than ever to address the complex challenges in increasing on-farm water use efficiency.

Note

1   Details of the 2008 Proposition 50 Proposal Solicitation Package is available on the DWR website, www.owue.water.ca.gov/finance/index.cfm.

References

AWMC (1999) *Memorandum of Understanding Regarding Efficient Water Management Practices by Agricultural Water Suppliers in California*, Agricultural Water Management Council, Agricultural Water Suppliers, Efficient Water Management Practices Act of 1990: Assembly Bill 3616. Sacramento, CA

Davidoff, B., Craddock, E., Roose, M. and Karajeh, F. (1996) 'Optimum on-farm irrigation efficiency for sustainable agriculture', in *North American Water and Environment Congress 1996*, ed. Chenchayya Bathala, American Society of Civil Engineering, New York, pp189–194

Department of Water Resources (1994) *California Water Plan Update*, Bulletin 160-93, DWR, Sacramento, CA

Department of Water Resources (1997) *Fifteen Years of Growth and a Promising Future*, CIMIS, The California Irrigation Management Information System, DWR, Sacramento, CA

Department of Water Resources (1998) *Technical Elements of CIMIS, The California Irrigation Management Information System*, The Resource Agency, Division of Planning and Local Assistance, DWR, Sacramento, CA

Department of Water Resources (2005) *California Water Plan Update: A Framework for Action October 1994*, Bulletin 160-05, DWR, Sacramento, CA

Hoagland, R. and Davidoff, B. (1998) 'On-farm agricultural water management', in E. Cabera, R. Cobacho and J. Lund (eds) *Regional Water System Management: Water Conservation, Water Supply, and System Integration*, A.A. Balkema Publishers, the Netherlands, pp93–112

Hockett, B. (2006) *Annual Report: North Kern Resource Conservation District*, North West Kern Resource Conservation District, Bakersfield, CA

Parker, D., Cohen-Vogel, D. R., Osgood, D. E. and Zilberman, D. (2000) 'Publicly funded weather database benefits users statewide', *California Agriculture*, vol 54, May–June

Pitts, D., Peterson, K., Gilbert, G. and Fastenau, R. (1996) 'Field assessment of irrigation system performance', *American Society of Agricultural Engineers*, vol 12, no 3, pp307–313

Sanden, B., Poole, G. and Hanson, B. (2003) 'Soil moisture monitoring in alfalfa: Does it pay?', in *Proceedings 33rd California Alfalfa Symposium*, 17–19 December 2003, Monterey, CA, California Alfalfa and Forage Systems Workgroup, UCCE and Department of Plant Sciences, Davis, CA, pp53–65, http://ucanr.org/alf_symp/2003/03-53.pdf

Solomon, K. H. and Davidoff, B. (1999) 'Relating unit and sub-unit irrigation performance', *American Society of Agricultural Engineers*, vol 42, no 1, pp115–22.

United Nations (1999) *The World at Six Billion*, Department of Economic and Social Affairs, Population Division, UN Secretariat, available at www.un.org/esa/population/publications/sixbillion/sixbillion.htm

10

Pricing Savings, Valuing Losses and Measuring Costs: Do We Really Know How to Talk about Improved Water Management?

Chris Perry

An economist might be described as someone who doesn't see anything special about water. (Tregarthen, 1983)

Introduction

Irrigation uses more water than any other sector – especially in water-scarce countries with erratic or limited rainfall. In many locations, irrigation from groundwater sources exceeds the safe yield of aquifers, silently and invisibly compromising the future. Often the water is used to produce low-value crops and far more water is applied to fields than is warranted by scientific analysis. Meanwhile, towns and cities, industries, tourist facilities and ecosystems are short of water.

Unsurprisingly, irrigation is widely accused of being a wasteful, low-value use of water. Pressure mounts to restrict its share of water use – to limit abstractions to sustainable levels and transfer water to more socially valued purposes.

Interventions to address the problem reflect different perspectives – social, political, technical and economic. Technical and economic proposals are often the most specific – 'save water through improved irrigation technology', 'introduce water markets or volumetric charging'. Some also suggest to 'limit the water allocated to irrigation' – i.e., introduce quotas – and come from relatively well-articulated sources (Bos and Walters, 1990; World Bank, 1993; Ward, 2000; Burt, 2002). However, the implications of the proposed interventions are often poorly understood because the language for the analysis is opaque and open to multiple interpretations.

In the section entitled 'Water use, consumption, losses and efficiency', recent work on the technical definitions that underpin the analysis of physical interventions is presented, showing that there is considerable scope for confusion in the traditional terminology – efficiency, savings, losses – especially as one moves from on-farm water management to basin-scale resource management. A set of terms recently adopted by the International Commission on Irrigation and Drainage (ICID) are described and justified.

In the next section, 'Economic instruments', attention is paid to the terminology of economic incentives – prices, costs, values – and clarification of these is presented, again based on recent literature.

Finally, these two sets of terminology are related to each other, identifying which constructively interact with which, and which are unrelated and as such not useful as explanatory relationships.

This chapter does not present ways to save water or indeed to manage it better. It is designed to help the people who understand how to undertake these critical tasks to explain the rationale for their recommendations (and the irrationality of some of the solutions proposed by others).

Water use, consumption, losses and efficiency: A brief history and recent thinking

Engineering considerations have historically dominated the approach to water accounting. Irrigation facilities (diversion weirs, dams, canals, pumps, etc.) are sized with reference to the availability of, and need for water, and the proposed area for irrigation is sized to match the supply of water at the field to the demand for water of the proposed crops under the local climatic conditions. Economy of design in irrigation and other water-using sectors require that expensive facilities should be of the minimum size necessary, and that as much as possible of the water that the facilities have stored, diverted or pumped should reach the intended productive purpose of supporting crop transpiration or any other intended use. In irrigation, much attention in consequence has been paid to the ratios between the volume of water available at the diversion point or storage reservoir, the volume of water delivered, successively, to the farm, field and crop, and the volume of water utilized productively by the crop.

The concept of efficiency in irrigation evolved some 60 years ago. Following extensive fieldwork in the 1940s, measuring the quantities of water applied to fields compared with the actual evapotranspiration requirement, Israelsen (1950) stated: 'With a given quantity of water diverted from a river, the larger the proportion that is stored in the root-zone soil of the irrigated farms and held there until absorbed by plants and transpired by them, the larger will be the total crop yield.' He then defined irrigation efficiency as the ratio of the irrigation water consumed by the crops of an irrigated farm or scheme to the water diverted from a river or other natural water source into the farm or scheme canal or canals.

Essentially, he defined irrigation efficiency as the ratio of water consumed by the intended purpose to that diverted. This approach to irrigation accounting remained fundamentally unchanged for more than 40 years. Reservations and refinements were suggested – for example, Hansen (1960) pointed out that if the water applied is less than the potential consumption by the crop, the water application efficiency may approach 100 per cent, but the irrigation practice may be poor and the crop yield low – so that high efficiency was not reliably correlated with good performance. He proposed to disaggregate efficiency into a number of components and proposed an overall concept of consumptive use efficiency.

Jensen (1967) pointed out that for sustained irrigated agriculture, the quantity of water effectively used to control soil salinity (the leaching fraction) should be considered a beneficial use. Therefore, he defined irrigation efficiency as the ratio of evapotranspiration (ET) of irrigation water plus the water 'necessary' for leaching on a steady-state basis to the volume of water diverted, stored or pumped specifically for irrigation. Subsequently, Jensen (2002) has pointed out that this resulted in the numerator containing a consumptive component and a small non-consumptive component, making water balance calculations more complex. Bos and Nugteren (1974, 1982) published the results of a joint effort of the ICID, the University of Agriculture, Wageningen, and the International Institute for Land Reclamation and Improvement (ILRI), Wageningen. The definitions of efficiency terms were refined in the 1982 edition. Distribution efficiency was defined as the ratio of the volume of water furnished to the fields, to the volume of water delivered to the distribution system. Field application efficiency was defined as the ratio of the volume of irrigation water needed, and made available, for ET by the crop to avoid undesirable water stress in the plants throughout the growing cycle, to the volume of water furnished to the fields. Combining these various figures at appropriate scales provided measures of efficiency at field, farm, tertiary, scheme and district level.

Despite these variations and enhancements, Israelsen's original definition of efficiency, relating the water used by the crop to the water diverted at some point, remained the underlying accounting basis in irrigation. Since the various losses (in distribution and field application) were essential knowledge to those designing the irrigation systems because the delivery quantity at successively higher levels needed to provide for the water that was lost, this accounting basis was appropriate and relevant to that engineering purpose. High efficiency implied that a high proportion of the water available at the head of a scheme was being used for the design purpose of augmenting crop transpiration – an appropriate engineering objective.

However, as demand for water from various sectors has increased and supplies have become relatively scarce, the narrow and local 'irrigation engineering' definition of water accounting has become less useful and indeed prone to produce misleading indicators. When water in a basin is scarce, the impacts of one use are felt by others. River basins are naturally 'integrating' entities: the observed status at any point is the sum total of whatever happened above – rainfall, storage, releases, ET – indeed the original notion of integrated water resources management had far more basis in these physical interdependencies than the set of issues now collected under that simple umbrella.[1]

At the wider scale of basin analysis, clear distinctions must be made between consumptive uses, which remove water from the current hydrological cycle, and non-consumptive uses of water, which return the water for potential reuse. This distinction is not relevant to the irrigation engineering of Israelsen, but becomes essential as scarcity and different types of use begin to interact.

For example, in a typical house connected to a main sewer system, some 95 per cent of the water delivered by the water utility is collected and returned for treatment and subsequent reuse within the water resources system. But wild claims are made in respected journals (*Scientific American*, 2001) that vast quantities of water can be 'saved' by increasing 'efficiency' through the use of low-flow showers and mini-flush toilets. In fact, the *consumptive* use of a shower, like a bath or a toilet, is essentially zero if it is connected to a sewer system. It is the hydrological location of the diversion and return flows that determines the impact of such uses on the availability of water for alternative uses. It is often said that Londoners are the fifth people to drink the water of the Thames – and everyone understands the implications! – but in parallel with this intuitive understanding of water use and reuse many find it hard to understand that 'using less' water may not 'save' water, and homeowners upstream in the Thames valley are provided with subsidized barrels to 'save' rainwater.

This issue was highlighted in a recent exchange on the WinrockWater website. Peter Gleick submitted two notes (Gleick, 2006a, b) as a basis for explaining how the introduction of low-flow toilets saved water.

The first note – a memorandum – summarizes the results of an analysis of the impact of introducing low-flow toilets on the 'urban demand' for water. The analysis demonstrates substantial reductions in something called 'real demand' for water and accuses the California Department of Water Resources of failing to account properly for this potential.

The associated document is a hydrologic flow chart showing diversions to two cities for various uses, return flows to the river system and the final outflow to the ocean after the cities have taken their requirements in accordance with normal and 'more efficient' needs. Surprisingly, the difference in final outflow between the original case and the case after improved, efficient, 'real demand'-reducing toilets are installed is zero. So a downstream user, or an ecosystem hoping to find more water available following the introduction of more 'efficient' infrastructure upstream, would be disappointed. In fact, the analysis could have made the very important point that if the return flows from the toilets were not recovered through the sewage system for downstream use, then the 'savings' would indeed have been observable as increased downstream availability: hydrological context is all important and masked by the terminology. Gleick claims quite rightly that low-flow toilets reduce water diversions by cities, reducing costs for water treatment, reducing the need for upstream water storage to sustain supplies during droughts and reducing local dewatering of streams between points of diversion and points of return. However, claims implying the most obvious 'saving' – of physical, wet, fungible water – are misleading and, indeed, usually wrong.

Such confusion can be observed in authoritative data sets. The Pacific Institute (2007) quotes Egypt's annual renewable water resource as $86.8km^3$ – a surprisingly high figure, given that Egypt's agreed share of the Nile is $55.5km^3$ and rainfall is negligible.

Meanwhile, Earthtrends (2007) reports a figure of $58km^3$ with 'internal renewable resources adding an additional 1 cubic kilometer'. Both sources refer to the Food and Agriculture Organization's AQUASTAT as the basis of their information.

The fact that estimates of the available water in the most regulated and documented large water system in the world should vary by some 50 per cent must be a cause for concern.

Getting the terminology right, so that irrigation engineers, water supply and sanitation engineers, hydrologists, planners and journals can all contribute meaningfully to an important debate is a high priority. To this end, the ICID community has undertaken an international consultation (Perry, 2007) aimed at achieving consensus on this issue.

Others are already working in this direction. The draft water requirements chapter of the upcoming revision of the American Society of Agricultural Engineers' (ASAE) *Monograph on Design and Operation of Irrigation Systems*, and the upcoming revision of the American Society of Civil Engineers' (ASCE) *Manual 70 (Evapotranspiration and Irrigation Water Requirements)* are expected to replace 'efficiency' terms with alternative terminology that reduces the scope for confusion and misuse.

This intitiative (or at least its analytical underpinning) is not entirely novel. In 1979, a US Interagency Task Force completed a report, *Irrigation Water Use and Management* (US Interagency Task Force, 1979). The task force based its report on available literature and input from a number of specialists, and undertook a detailed review of field data on irrigation efficiency and related information on water laws and institutions, causes of inefficiencies and their results. Regarding irrigation efficiency, it stated: 'Any report dealing with irrigation efficiencies must first define "efficiency" with a great deal of care. Many different and sometimes conflicting definitions have been published. It is frequently assumed that because irrigation efficiency is low, much irrigation water is wasted. This is not necessarily so' (p22). This significant change in thinking was endorsed by Jensen (1993) who referred to water balance and river basin studies, providing an important intellectual link, as the 1979 task force had done, to the holistic approach of the hydrologist.

In fact, this link is critical: hydrology studies water flows at basin scale, making no value-laden distinctions about 'losses', 'waste' or 'efficiency'.[2] Rather, the water flows are traced from source to use, or sink.

The next 10 years saw a number of important contributions to the 'irrigation efficiency' debate – Willardson et al (1994), Allen et al (1997) and Willardson and Allen (1998) suggested that the 'classical' efficiency term was outmoded. This series of papers recommended using ratios or fractions to describe water use and to explicitly consider impacts of return flows. Perhaps equally important was the move away from the value-laden term 'efficiency'.

Non-engineers have added to the literature on the subject of irrigation efficiency. That of Seckler (1993) provided the foundation used in the development of the new International Water Management Institute (IWMI) framework for water resources analysis, referred to as the 'IWMI Paradigm' which analyses irrigation water use in the context of the water balance of the river basin (Perry, 1999). Others have documented similar views in the past – indeed, all hydrological models incorporate much of this logic as a matter of course.

Molden (1997) developed procedures for accounting for water use, or water accounting based on a water balance approach. Water accounting is a procedure for analysing the uses, depletion and productivity of water in a water basin context. A key term is water depletion, which is the use or removal of water from a water basin such that it is permanently unavailable for further use. He described process and non-process depletions. Process depletion is where water is depleted to produce an intended good. In agriculture, process depletion is transpiration plus that incorporated into plant tissues – the product. Non-process depletion includes evaporation from soil and water surfaces and any non-evaporated component that does not return to the freshwater resource. The depleted fraction is that part of inflow that is depleted by both process and non-process uses of water. While the term 'depletion' allows the amalgamation of all the components that remove water from the renewable water resource system (evaporation, transpiration, flows to sinks and pollution), it does not conform to the more general meaning of the term, which implies removal from storage (e.g. depletion of an aquifer or reservoir).

Molden further suggests that the productivity of water can be measured against gross or net inflow, depleted water, process-depleted water or available water in contrast to the production per unit of water consumed in ET. Water accounting can be done at various levels such as the field, irrigation service, basin or sub-basin levels. Molden and Sakthivadivel (1999) presented a detailed example of water accounting at the basin level using data from Egypt's Nile River where detailed information on water use and productivity was available. This study made clear how the computed 'classical' efficiency of irrigation varied substantially with scale: measured at the basin level, Egyptian irrigation is approximately twice as efficient as it is at field scale. Molden and Sakthivadival (1999) presented another example for a district in Sri Lanka.

The general thrust of these papers was to divide water diverted to irrigation schemes into various fractions. After lengthy internal discussions, the ICID has adopted the following definitions:

- Water use: any deliberate application of water to a specified purpose. The term does not distinguish between uses that remove the water from further use (evaporation, transpiration, flows to sinks) and uses that have little quantitative impact on water availability (navigation, hydropower, most domestic uses).
- Withdrawal: water abstracted from streams, groundwater or storage for any use – irrigation, domestic water supply, etc. Within withdrawals, following the recommendations of Willardson et al (1994) and Allen et al (1997), water would go to:
  1 Changes in storage (positive or negative): changes in storage include any flows to or from aquifers, in-system tanks, reservoirs, etc. The key characteristic of storage is that the water entering and leaving is essentially of the same quality.
  2 Consumed fraction (evaporation and transpiration) comprising: first, beneficial consumption, i.e., water evaporated or transpired for the intended purpose, e.g. evaporation from a cooling tower, or transpiration from an irrigated crop; and second, non-beneficial consumption, i.e., water evaporated or transpired for purposes other than the intended use, e.g. evaporation from water surfaces, riparian vegetation or water-logged land.
  3 Non-consumed fraction, comprising: first, recoverable fraction, i.e., water that can be captured and reused, e.g. flows to drains that return to the river system and percolation from irrigated fields to aquifers, and return flows from sewage systems; second, non-recoverable fraction, i.e., water that is lost to further use, e.g. flows to saline groundwater sinks, deep aquifers that are not economically exploitable, or flows to the sea. This framework is consistent with hydrology – it meets the criterion of continuity of mass. It recognizes that water is water – it is not somehow differentiated by colour or source.

Within this framework it would be clear that the key areas of attention when water is scarce would be to reduce non-beneficial consumption, and to reduce non-recoverable flows to the extent that proper hydrological analysis shows that no unintended consequences of such reductions occur.

This set of ideas adds importantly to Israelsen's original analytical framework: first and most importantly, it provides a common terminology for all water-using sectors to communicate with each other; second, within the 'consumptive' sectors, it focuses attention on what is really a loss (non-beneficial ET, and the non-recoverable component of the non-consumed fraction).

For completeness, two other issues were clarified in the consultation process within ICID which are relevant to discussions of irrigation and water resources management: first, that the term 'water use efficiency' should be replaced by 'productivity of water';[3] and second, that the issue of pollution, while critical, could not be incorporated into the 'fractions' approach (or indeed any other quantitative analysis) because the significance of pollution depends on the intensity of pollution and the downstream use. Pollution must therefore be assessed in a separate analysis.

Economic instruments

If scarcity and competition for water are the primary concerns of practitioners, scarcity of the funds required to sustain the functionality of systems and expand services to new users is close behind. The issue is not confined to 'poor farmers' in 'developing countries'. Most observers agree that the privatization of water supply in the UK was driven by a political preference to keep investment in new facilities (and upgrading existing facilities to comply with the European Water Framework Directive) separate from the public sector borrowing requirement.

For an economist, the combination of an often unpriced scarce resource and shortage of funds to deliver a service point irresistibly towards market forces as a means to achieve the twin objectives of limiting demand and raising funds. The terminology that has grown up around this area is sometimes as vague as the 'efficiency' terminology addressed in the previous section.

'Economic instruments' are commonly proposed as a means to achieve 'demand management' and achieve 'optimal allocation' of water through 'water markets'.

What do these terms mean and do they help the policy debate?

First, economic instruments generally means using pricing to encourage water users to recognize that the resource they are using is (a) not unlimited and (b) costs money to deliver. Demand management means using economic instruments to reduce demand: in the water supply sector, increasing the cost of the service has a proven record of reducing demand for water. The term also has the particular role of distinguishing itself from supply management (i.e., increasing the availability of the resource to balance supply and demand). Optimal allocation is usually referred to in the context of alleged waste and low-value use in agriculture, and indicates that water should be allocated in order of priority to its highest value uses, and water markets are suggested as the means by which such allocation can take place.

Before returning to first principles and clarifying how economics in general and pricing in particular may contribute to water management in the context and terminology described in the previous section, a few general observations based on the author's experiences may be helpful.

In many areas of the world, water resources are over-used – aquifers are falling, wetlands are drying up, etc. This situation reflects a mixture of *de jure* policies to develop water, as well as *de facto* development without permission. Retrofitting 'sustainable use' to this situation is politically extremely hard. In fact, it means shutting down significant quantities of consumptive use and politicians are understandably reluctant to embark on that path until all other options have been exhausted (which in part explains the historical dominance of the 'efficiency' paradigm). 'Demand management' through 'economic instruments' is appealingly neutral and 'hands-off' when contrasted with forcing abandonment of wells or withdrawing historic irrigation entitlements.

Similarly, while politicians everywhere are reluctant to increase service charges to levels that meet even operation and maintenance (O&M) costs, let alone the full cost recovery objective in the Water Framework Directive, the idea that markets can improve cost recovery – again without identifiable interventions by politicians – is also attractive.

In reality, of course, market outcomes may not be at all what users, politicians or society at large would like. The rhetoric that accompanies the promotion of demand management, economic instruments and optimization is also about gender equality, intergenerational equity and pro-poor interventions. None of these laudable objectives is likely to be served by 'markets'. Even where markets are thought to be serving the optimization function, a clearer look suggests that the analyst may be confused: much of the literature on groundwater markets rests on the erroneous assumption that the market is indeed for water, but since all suppliers are drawing from the same aquifer, the market is in fact for pumping services, and the more water a service provider can pump at the lowest cost, then the happier the 'market analysts' will be and the quicker the aquifer will be pumped to extinction. Markets without defined, sustainable rights are dangerous. For further elaboration of these themes, see Perry et al (1997) and Hellegers and Perry (2004).

With these caveats in mind, we return now to first principles. What are we trying to achieve and what contribution can 'economic instruments' make to that goal?

The three most widely stated objectives of irrigation charging are:

1   to achieve a specified and consistent level of cost recovery from users;
2   to provide an incentive to irrigators to reduce water consumption (demand management);
3   to increase the productivity of water at the individual user level, or through transfers to more productive users or uses, with the objective of increasing the level of economic benefit per unit of water.

The detailed interpretation and realization of these objectives is not simple. 'Cost recovery' can range from simple day-to-day operational costs to the entire cost of operation, maintenance, future replacement and amortization of past investments. Full cost recovery is fraught with complexity: if a water supply system was built more than 100 years ago by a private company that went bankrupt, what residual 'capital costs' remain to be recovered? (This fanciful sounding example is actually the Punjab.) In fact, when the UK water companies were privatized, historic capital costs were essentially written off.

The terms charge, price, cost and value are commonly used interchangeably. To avoid confusion, the terminology recommended here is based on the following definitions (Cornish et al, 2004):

- Irrigation service charge: the total payment made by a user for an irrigation service. It may comprise fixed elements (e.g. US$20/ha) plus variable elements (e.g. US$1/1000m^3 water). In this example, if a user with one hectare took 10,000m^3 under the above charging system, the charge would be US$30.
- Average price: in the above example, the average price of water would be the total charge divided by the total quantity of water received: (US$30/10,000m^3 = US$0.03/m^3).
- Marginal price: the cost of an additional unit of water: (US$1/1000m^3 = US$0.001/m^3)
- Cost of the irrigation service: the expenses incurred by the supplying agency in providing the service. Precise definitions depend on local rules, but typically include operation, maintenance, staff and fuel costs, plus some elements of replacement costs and amortization of capital.
- Value of water: incremental income received by the farmer as a result of irrigation services, divided by the quantity of irrigation water used. (Value may also be distinguished as average or marginal, but data are rarely available to do so.)

The most widely practised systems of charging for irrigation service are area-based charges, crop-based charges and volumetric charges. For completeness, and because each is a form of economic instrument or at least creates an economic response, rationing (quotas) and tradable water rights are also considered.

Area-based charges are commonly applied in mono-crop areas, where farmers tend to plant and irrigate their entire area: thus in some rice growing areas of eastern India, farmers are charged in proportion to the area they own unless they can demonstrate that they did not receive water. More commonly, an area-based charge is one component of a two-part tariff, with the second part either volumetric or quasi-volumetric. This provides agencies with a higher degree of stability in their income.[4]

More commonly, farmers are charged on the basis of the area irrigated and type of crop. This allows for considerable complexity in charging: for example, charges can be set to reflect the water consumption of the crop (and hence be quasi-volumetric) or set to charge higher rates for cash crops and lower rates for subsistence crops. Crops that are seen as having strategic importance can be favoured, and so on.

Volumetric charges are based on the volume of water delivered to the farm and must be based on actual measurements acceptable to both the farmer and the agency. The charge for a volumetric service can be on various bases – for example, a simple '$/m^3' price or a tiered pricing structure (basic allowance at a low price per m^3; with subsequent tranche(s) at progressively higher prices). It is also important to identify whether the service actually provides as much water as the user demands at the agreed price (which is rare) and whether the service charges volumetrically, but specifies the maximum allowance of water (usually m^3/ha – for example, as practised in Morocco (Hellegers and Perry, 2004)). Indeed, tiered pricing structures usually set the final price so high as to approximate a quota system. Other variations include the 'warimetric' system which was experimented with in northern India some years ago, where the irrigation charge was related to the number of turns received during the irrigation season. As the farmers had no choice in how many turns they received, there was no 'incentive' in this system, rather an *ex post* attempt at equity.

More recently, the possibility to remotely measure ET from irrigated areas has provided the potential to actually charge on the basis of water consumption (a step that greatly improves the clarity of incentives, as will be discussed later).

With well-defined water rights in place,[5] it is possible to consider the scope for tradable water rights. These have the benefit of separating the charge for O&M that agencies need for financial sustainability from the pricing of the resource. Also, since the resource price is a 'shadow' price, the farmers are faced with the incentive to utilize water productively themselves or to save water and sell surpluses to a more productive user, but they are not faced with the equivalent actual financial price, which would render many uncompetitive. On the other hand, the potential third party impacts of transfers mean that an expensive administrative approval and record-keeping structure is required.

Economic instruments and water accounting

The two previous sections of this chapter have set out terminology in the distinct areas of physical water resources management, and the economic instruments available to encourage better water use. This section attempts to demonstrate how interventions in these two dimensions interact, beginning with a detailed analysis of a particular case, followed by a summary of significant interactions for the array of situations identified above.

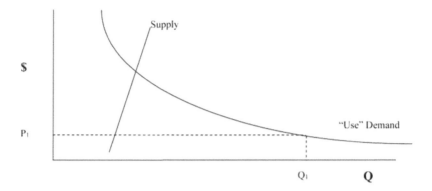

Figure 10.1 *Demand, supply and price of water in a typical irrigation system*

Figure 10.1 shows a very common situation in irrigation systems. The price of water is far below that required to balance supply and demand and, for most farmers, substantially below the value they derive from irrigation. In such circumstances, farmers' behaviour is driven not by the price of water, but by its value. Each will try to obtain as much water as possible – often at the expense of less well-situated farmers. Faced with this situation, planners and policy makers wishing to bring supply and demand into balance may consider pricing and the first question to ask is whether the price increase required for equilibrium is within the 'politically feasible' envelope. In the graph, the increase would be 250–300 per cent.

Often the ratio is much higher. In a study of five irrigated areas in various countries (Hellegers and Perry, 2004; see Table 10.1), the ratio between the price paid for water (computed as total payment divided by volume received, regardless of the structure of charges) and the cost of delivery (O&M only) varied from 0.8 in Morocco (where somewhat more than O&M charges are recovered) to 25 in Egypt. Comparing price with value gave a ratio of 5 in Morocco, indicating that even where the price is high, it is value that dominates farmers' behaviour, up to 200 in Egypt and Indonesia. Another interesting aspect of these data is that there appears to be a relationship between the cost of the service and the value achieved.

Table 10.1 *Price, cost and value of water ($ / m³) in selected irrigation areas*

	Price paid ($/m³)	Price/cost ratio	O&M cost ($/m³)	Price/value ratio	Value of water ($/m³)
Kemry (Egypt)	0.0004	1:25	0.010	1:200	0.08
Haryana (India)	0.0005	1:2.6	0.0013	1:80	0.04
Tadla (Morocco)	0.0200	1:0.8	0.017	1:5	0.10
Brantas (Indonesia)	0.0002	1:5.0	0.001	1:200	0.04
Crimea (Ukraine)	0.0020	1:6.0	0.012	1:55	0.11

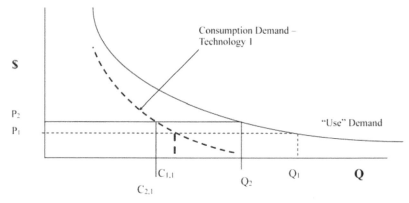

Figure 10.2 *Consumption demand and use demand*

A reasonable conclusion that can be drawn from these data is that while proper financing for the *cost* of water services is essential for the sustainability of services, the *price* required to meet that objective is unlikely to have much impact on demand for water, consumption or productivity.

Figure 10.2 illustrates the distinction between water used and water consumed. In the graph, the consumed fraction is about half of the water applied to the field – corresponding to an 'irrigation efficiency' of 50 per cent. The two curves are important: the 'use demand' curve is the conventional basis for volumetric charging; and the 'consumption demand' curve is the actual water consumed by the crop – which is the part of the water use that provides benefit to the farmer. An increase in the price of water from P_1 to P_2 induces a fall in use demand and consumptive demand from Q_1 to Q_2 and $C_{1,1}$ to $C_{2,1}$, respectively. (The second subscript indicates the technology.) So far, all is as expected: an increase in the price of an input has induced a fall in demand.

In Figure 10.3 the effect of a change in technology is explored: Technology 2 is introduced, with a consumed fraction of around 75 per cent. The implication of this change in the consumed fraction is that for every unit of 'use', the farmer now gets an additional 50 per cent (75/50 = 1.5) of consumptive, productive benefit. Such a shift in technology can be triggered by an increase in the price of water or reduced availability of water. The result is a sharp increase in the productivity of 'used' water and, if the farmer has any additional land, the likely result of this technology shift will be as indicated by the new consumption level $C_{2,2}$. The significant point here is that the new consumptive demand is *higher* than the original demand – in other words, an increase in price of water *used* has prompted a technology shift and resulted in an increase in water consumption. Not the anticipated result and probably not the desired result.

Of course, how a shift in price will actually play out depends on many factors and the example here is hypothetical, but feasible – indeed it seems to be exactly the result that the European Water Framework Directive (seeking higher

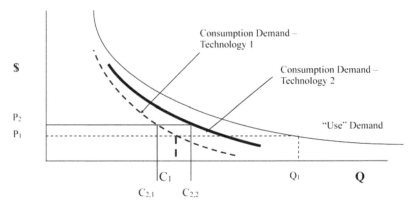

Figure 10.3 *Consumption demand and use demand – higher prices and improved technology*

water prices, closer to full economic costs and encouraging 'efficiency') is having on aquifers in Spain.

Table 10.2 sets out as an array the economic instruments and the water fractions identified above. Each intersection is worthy of some discussion, but here only a few key relationships are noted.

First, a fundamental distinction is made between those instruments that are usually based on the cost of providing the service. As noted above, in many systems costs are not recovered and, where they are, the price required to meet the cost recovery objective is generally modest in relation to the value of the water to the user. Value-related instruments, on the other hand, expose the user to the shadow price of the resource – either the income forgone by misuse or waste, in the case of a quota, or the income forgone by not selling the water, in the case of tradable water rights. As shown in Table 10.1, and in many other studies, the value of water is generally far higher than the price. In consequence, the response to a value-related instrument is likely to be far sharper than the response to a cost-related instrument.

Generally, area-based charges and crop-based charges have negligible impacts on the component fractions of water use. Only severely skewed charges

Table 10.2 *Economic instruments and water fractions*

	Area-based	Cost-related instruments Crop-based	Volumetric	Value-related instruments Quota	Tradable
Consumed		↓	↓↑ ?	↑↑	↑↑
Beneficial			↑	↑↑↑	↑↑↑
Non-beneficial			↓	↓	↓
Non-consumed			↓↓↓	↓↓↓	↓↓↓
Recoverable					
Non-recoverable					

favouring crops with high water consumption and low water productivity (sugar cane, for example) may have a small impact on consumption.

As examined in more detail above, volumetric charges may reduce or increase water consumption depending on the technical response of farmers. In general, the non-consumed fraction should reduce as volumetric charges increase because an increase in the consumed fraction incurs no incremental volumetric charge (or indeed a reduction, if consumptive use remains constant).

The strongest relationships are to the value-related instruments. Faced with a high shadow value for water, farmers will seek to increase the consumed fraction, and especially beneficial consumption, precisely because the shadow value is the result of beneficial consumption, not use. Switching to quotas or tradable water rights based on historic diversion entitlements is thus likely to induce increased consumption and result in less water available downstream or for aquifer recharge. (This seems to have happened in the Murray–Darling basin in Australia, where the specification of rights and introduction of trading suddenly made unused rights valuable in the market. They were sold to users and consumption increased.[6])

While policy makers and planners should be aware of the distinction between recoverable and non-recoverable flows, it is clear that farmers will be indifferent to whether the non-consumed fraction is recoverable or not.

A primary problem underlying the uncertain linkage between economic instruments and the actual 'water outcome' is that economic instruments are applied to 'use', which is not necessarily the outcome we wish to influence. If the objective is to increase water available in estuaries or wetlands, or to reduce over-abstraction from an aquifer, then consumptive use is the primary parameter of interest. Interestingly, new technologies are emerging that allow direct measurement of evapotranspiration, the dominant consumptive use.

Remote sensing provides information that can be interpreted spatially and temporally to map evapotranspiration. Pilot studies are under way to implement water rights systems that define entitlements in terms of consumption rather than abstraction. Under such systems, farmers have incentives to minimize non-beneficial evapotranspiration and maximize the productivity of beneficial use – exactly as policy makers would wish. However, since there would be no incentive to minimize non-recoverable losses, the level of water use will still remain of interest to water managers and an appropriate target for incentives.

Conclusions and policy recommendations

The dominant framework for analysing water use in irrigation projects has been the engineering perspective, which focuses on 'efficiency' at various scales and emphasizes the need to avoid 'losses'. As water has become scarce at the basin scale, this narrow framework can lead to erroneous conclusions, most especially because many losses are recaptured for use elsewhere or at a later date. Equally importantly, some uses (irrigation, cooling towers) have the purpose of removing water from the hydrological cycle through transpiration and evaporation, while other uses (domestic, most industrial, navigation, hydropower) use water non-consumptively, returning virtually all to the system for further use. Understanding the hydrological implications of these differences is critical to sound water policy.

The ICID has recently adopted a new set of terms that avoid this problem, categorizing use as consumptive or non-consumptive fractions that are either recoverable or non-recoverable and beneficial or non-beneficial. This framework allows planners to focus on those parts of the hydrology of a basin that are a priority for improving water availability – reducing non-beneficial consumption and reducing non-recoverable flows. Further, this allows various sectoral 'uses' to be assessed within a common framework.

An important conclusion from the revised analytical framework is that 'improved' irrigation technologies certainly increase the *consumption* of water and as such may have negative impacts on the resource balance. Groundwater recharge is likely to fall, return flows to rivers are likely to be reduced and ecosystem balances will become more precarious.

As concerns over physical availability of water have grown, the potential for the application of economic instruments (pricing, markets, rationing) has attracted increasing attention. However, the terminology of this debate has frequently failed to distinguish the basic differences between prices, values and costs, and unless these concepts are fully clarified, it is impossible to assess meaningfully the impact of an economic intervention, e.g. raising the price of water, either on farmer behaviour or on the water balance.

In many cases, an increase in the price of water encourages investment by the user to increase the consumed fraction. The introduction of markets exposes users to those who are most short of water, and best able to pay for it – those most keen to increase consumption. Neither prices nor markets are thus likely to decrease the consumption of water

In considering any intervention to change the behaviour of water users, it is always essential to fully understand the hydrological context – where is the water currently going? which aspects of that flow pattern are we trying to change? – and understand what signal the user is reacting to – the price of water, or its value.

Notes

1   The much referenced Dublin Principles (fresh water is a finite and vulnerable resource; water development and management should be based on a participatory approach; women play a central part in the provision, management and safeguarding of water; and water should be recognized as an economic good) are a worthwhile political statement, but hardly helpful to the analysis and physical management of the resource.

2 And hydrology recognizes too that water is water – not colour-coded into blue, green or grey as if one colour can be used independently without impact on the others.

3 While water use efficiency is clearly and precisely defined in various places, it remains one of the most misused terms in the literature, often being confused with the traditional irrigation efficiency term. For classical confusion, see *Water Use Efficiency at the River Basin Scale: Implications for Hydrologic Science and Water Management Policies*, Ximing Cai, International Food Policy Research Institute (IFPRI), International Water Management Institute (IWMI), 2033 K St, Washington, DC, 2006. The paper aims to clarify the confusion that may occur when field-scale terminology is applied to basin level, and in doing so misuses the standard term for productivity and relates the discussion to 'hydro-logical science'!

4 While beyond the scope of this paper, the issue of income stability for irrigation agencies is important. It is addressed more fully in Cornish et al (2004).

5 Something that is far easier written than achieved. It took the Murray–Darling system almost 20 years to define water rights adequately to provide the basis for initial trading – and they began from a well-defined system with respect for the rule of law, educated farmers, an informed press and political establishment (Don Blackmore, personal communication).

6 Don Blackmore, personal communication.

References

Allen, R. G., Willardson, L. S. and Frederiksen, H. (1997) 'Water use definitions and their use for assessing the impacts of water conservation', in J. M. de Jager, L. P. Vermes, R. Ragab (eds) *Proceedings of the ICID Workshop on Sustainable Irrigation in Areas of Water Scarcity and Drought*, Oxford, pp72–82

Bos, M. G. and Nugteren, J. (1974) *On Irrigation Efficiencies*, 1st edn, International Institute for Land Reclamation and Improvement, Wageningen, The Netherlands, pp95

Bos, M. G. and Nugteren, J. (1982) *On Irrigation Efficiencies*, 3rd edn, International Institute for Land Reclamation and Improvement, Wageningen, The Netherlands, pp142

Bos, M. G. and Walters, W. (1990) 'Water charges and irrigation efficiencies', *Irrigation and Drainage Systems*, vol 4, pp267

Burt, C. M. (2002) 'Volumetric water pricing', Draft document prepared for Irrigation Institutions Window, World Bank

Cornish, G., Bosworth, B. and Perry, C. (2004) *Water Charging in Irrigated Agriculture: An Analysis of International Experience*, FAO Water Report 28, Food and Agriculture Organization, Rome

Earthtrends (2007) http://earthtrends.wri.org/searchable_db/index.php, accessed 1 May 2007

Gleick, P. (2006a) 'Urban water use projections', http://groups.yahoo.com/group/winrock-water/files/, accessed 9 August 2006

Gleick, P. (2006b) 'Demand management and water use efficiency', http://groups.yahoo.com/group/winrockwater/files/, accessed 9 August 2006

Hansen, V. E. (1960) 'New concepts in irrigation efficiency', *Transactions of the American Society of Agricultural Engineers*, vol 3, no 1, pp55–57, 61, 64

Hellegers, P. J. G. J. and Perry, C. J. (2004) *Water as an Economic Good in Irrigated Agriculture: Theory and Practice*, LEI, The Hague, The Netherlands

Israelsen, O. W. (1950) *Irrigation Principles and Practices*, John Wiley and Sons, New York, pp471

Jensen, M. E. (1967) 'Evaluating irrigation efficiency', *Journal of Irrigation and Drainage Division*, American Society of Civil Engineers, 93(IR1), pp83–98.

Jensen, M. E. (1993) 'Impacts of irrigation and drainage on the environment', 5th N. D. Gulhati Lecture, The Hague, The Netherlands, 8 September, French and English, 26pp

Jensen, M. E. (2002) 'Irrigation efficiency', *18th ICID Congress on Irrigation and Drainage*, Montreal, Canada

Molden, D. and Sakthivadivel, R. (1999) 'Water accounting to assess use and productivity of water', *International Journal of Water Resources Development*, vol 15, no 1, pp55–71

Molden, D. (1997) *Accounting for Water Use and Productivity*, IWMI/SWIM Paper No 1, International Water Management Institute, Colombo, Sri Lanka, pp25

Pacific Institute (2007) www.worldwater.org/data.html, accessed 1 May 2007

Perry, C. J. (1999) 'The IIMI paradigm: definitions and implications', *Agricultural Water Management*, April 1999, vol 40, no 1, pp45–50

Perry, C. J., Seckler, D. and Rock, M. (1997) 'Water as an economic good: A solution or a problem?' in M. Kay, T. Franks and L. Smith (eds) *Water: Economics, Management and Demand*, E&FN Spon, London

Perry, C. (2007) 'Efficient irrigation; inefficient communication; flawed recommendations', *Irrigation and Drainage*, Wiley Interscience, vol 56, pp367–378

Scientific American (2001) Special feature: 'Are we tapped out? Safeguarding every drop of water', February

Seckler, D. (1993) 'Designing water resource strategies for the twenty-first century', Discussion Paper 16, Center for Economic Policy Studies, Winrock International, Arlington, VA

Tregarthen, T. D. (1983) 'Water in Colorado, fear and loathing of the marketplace', in R. Anderson (ed) *Water Rights, Scarce Resources Allocation: Bureaucracy and the Environment*, Pacific Institute for Public Policy Research, pp119–136

US Interagency Task Force (1979) *Irrigation Water Use and Management*, US Government Printing Office, Washington, DC, pp143

Ward, C. (2000) 'The political economy of irrigation water pricing in Yemen', in A. Dinar (ed) *The Political Economy of Water Pricing Reforms*, Oxford University Press for the World Bank, pp381–394

Willardson, L. S. and Allen, R. G. (1998) 'Definitive basin water management', in J. I. Burns and S. S. Anderson (eds) *Proceedings of the 14th Technical Conference on Irrigation, Drainage and Flood Control*, USCID, 3–6 June, Phoenix, Arizona, pp117–26.

Willardson, L. S., Allen, R. G. and Frederiksen, H. (1994) 'Eliminating irrigation efficiencies', *USCID 13th Technical Conference*, Denver, Colorado, 19–22 October, pp15

World Bank (1993) *Water Resources Management*, A World Bank policy paper, Washington, DC

11

Institutional Factors and Technology Adoption in Irrigated Farming in Spain: Impacts on Water Consumption

Llorenç Avellà and Marta García-Mollá

Introduction

One of the most striking characteristics of irrigated agriculture in Spain is its enormous diversity, not only in terms of the usual structural variables such as farm size, plot distribution patterns, etc., but also in terms of geographical location (wet Spain and dry Spain) and crop type, hence the wide differences in climate conditions and available water resources. There are also considerable differences in unit water needs (measured in m^3/ha) and water productivity, both across crop types and geographical areas, and major disparities in added value between irrigated and rain-fed agriculture, especially in central and southern Spain.

Such variability of conditions in Spain's irrigated farmland suggests that factors predicting the adoption of water-conservation technology will differ according to the type of irrigation under consideration.

It is only recently that efforts have been attempted to persuade farmers, Spain's main water consumers, to save water. Indeed, the country has reformed its water policy from the supply model that predominated until at least the 1980s (Carles-Genovés, 2000), to a demand model in which water is still treated as a factor of production, but one that is scarce and requiring water-conservation technology. The transition from the supply to the demand model has been beset by inertia and is proving to be slow and patchy.

Public administration entities are currently diversifying their efforts by applying a wide range of policies. Spain's adoption of the European Union Water Framework Directive (the underlying philosophy of which is that water is an ecological and social good, rather than a factor of production), demand-oriented policies (the federal Irrigation Crash Plan and initiatives by states) and supply-oriented policies, based on the use of unconventional resources, designed to address current water shortages. This combination of water policies, nevertheless, contrasts with the existing climate of confrontation between political parties and public administrations, whose perspectives are based on reality-oversimplifying assumptions.

In Spain, the allocation of water resources and the adoption of water-conservation technologies are both heavily influenced by institutional factors, since the initiative and frequently the funding and/or subsidies for the adoption of water-conservation technology, especially in surface-water irrigation systems, often come from public administration agencies.

Changes in Spanish irrigation systems, moreover, especially conversion from flood irrigation to drip and sprinkler technologies and new investment in the distribution of water for channel and plot irrigation, have been taking place on a widespread basis over the past two decades.

This chapter analyses the underlying relationships between public and private economic issues and the adoption of water-conservation technology, making a distinction between plot consumption and consumption over larger areas – aquifer or unit of demand and hydrographic basins.

Following this introduction, the next section, entitled 'Public administrations and irrigation in Spain', analyses the institutional framework of the Spanish irrigation sector and the problems arising from power-sharing arrangements between the various public administration agencies for water management. In 'The diversity of irrigated farmland in Spain', the current diversity of Spanish irrigation systems is described and how this affects water consumption. The following section, 'The water-conservation technology adoption process and its impact on water consumption', discusses the present state of Spanish irrigation systems with respect to water-conservation technologies and analyses the main determinants of technology adoption and the results in terms of water savings. The chapter concludes with a series of reflections on these issues.

The research focuses mainly on two hydrographic basins, the Segura and the Júcar, characterized by an overwhelming specialization in irrigated fruit and vegetable production and by structural water deficit. This deficit is structural because water demand is well in excess of available supply under normal climatic conditions and without overdrafting aquifers.

Public administrations and irrigation in Spain

Institutional factors

A brief historical overview will help to understand the role of the institutions in water allocation and in the adoption of water-conservation technology, and also the current debate and regional confrontation surrounding water usage.

Conflicts over water usage date far back into history. In recent decades, however, they have taken on a regional dimension that far outreaches the control of age-old institutions such as the Valencia water court (Tribunal de las Aguas) or the South Eastern Alicante water courts (Tribunales Privativos de Aguas), which were set up to settle the sort of conflict that nowadays has to be settled by the Supreme Court or the Constitutional Tribunal.

The increase in water demand, together with technological advances permitting water storage and transport over long distances, accounts for the fact that control over water usage has taken on a territorial dimension well beyond the control of Spain's traditional institutions.

Spain's first Water Law, which was passed in 1879, gave public control over surface waters to the central administration and introduced a system of concessions for the allocation of collective and individual water rights. Hydrological planning covered all waters likely to be used for irrigation purposes except groundwater, which played a very minor role at the time the law was written. Once technological progress made it possible to abstract significant volumes of groundwater, an official permit (from the mining authorities) was required before it could be used for irrigation purposes. This law upheld existing rights, some of which predated the unification of Spain as a nation.

The regenerationist movement of the late 19th century (one of its most eminent figures being the Aragonese Joaquín Costa) extended the notion that the hydraulic infrastructure needed to enlarge the irrigation system, and thereby increase agricultural output and employment, could not be built by private initiative alone. Given its importance to the economic development of the country, the work had to be undertaken by the state. As a consequence, a large portion of the investment in hydro works was possible thanks to generous public subsidies, often with less than full cost recovery, and/or with the conversion of dry farmland and improvements in the irrigation system being declared a matter of national interest. In other words, the public authorities approved the transfer of resources from society as a whole to certain regions in the hope of recovering them in the form of taxes on the resulting wealth and income gains.

In the first third of the 20th century, water boards were established in each hydrographic basin to undertake the planning and construction of the hydraulic infrastructure and to supervise the granting of concessions. The priority in those days was to make water resources available to users in order to increase agricultural output. This gave rise to the notorious supply-oriented policy implemented by the state in response to the limited interest of private capital in the agricultural sector.

This supply-oriented policy enabled the construction of most of the reservoirs and major irrigation channels in existence today. It facilitated the delivery of the then plentiful resource to users and substantially extended the irrigation system (in terms of surface area and output) mainly as a means to increase exports to improve the balance of payments and promote employment in the receiving regions.

The supply-oriented model was based (and still is, according to some authors' appraisals of the National Hydrological Plan currently in force) on understanding water as a non-exhaustible factor of production, in which supply is plentiful relative to demand. The aim, therefore, is to address the 'poor' spatial and temporal distribution of water resources; and the state is expected to place water at the disposal of the user. The major milestones achieved under this model include practically the whole of the hydraulic policy applied up until the mid-1980s (all the reservoirs, the Tajo–Segura transfer, etc.). At this stage, the water policy paid little or no attention to issues such as water conservation (through management improvements or the adoption of water-conservation technology), water reuse, desalination, etc.

During the 1980s, population concentration into urban centres and the structural change in the economy that led to the development of the secondary and tertiary sectors and the decline of the agricultural sector, gave rise to the first inter-sectoral and inter-regional conflicts over the use of water, which, due to increased demand, had ceased to be viewed as a plentiful resource. It was then that the supply-oriented policy was first challenged (more on theoretical than practical grounds) and calls for a demand-oriented policy, focused primarily on water-conservation measures, began to be heard. This drastic change, predictably, came up against problems due to decades of inertia in both the public and private water management and water usage policies. As in the supply-oriented model, however, it was the public sector that was expected to promote, execute and subsidize the lion's share of the investment required to make water conservation a reality.

That era also saw the approval of the Spanish Constitution (1978) and the regional autonomy statutes leading to the devolution of state powers to the states (known as autonomous communities) giving them power of decision in areas such as the filtering and reuse of wastewater, the extension of irrigated farmland, agriculture, environmental policy and territorial management, which directly affected hydrological planning (in the so-called inland basins) and water cycle management.[1]

The implementation of the Spanish Constitution through the distribution of power between the federation, states and municipalities sparked off serious conflicts between the various public water cycle management institutions over the availability of water for irrigation purposes, arising from the changeover from a centralized to a decentralized or mixed management regime.

Prior to the development of the state's autonomy, the federation had held ultimate control over water and agricultural management, which benefited coordination, despite the executive responsibilities falling to different government ministries. The central government was therefore responsible for water policies and for allocating resources in such a way as to balance supply with demand.

It is only fair to acknowledge that there were periods of major supply/demand imbalances even under central government control. The case of the Segura basin and the Tajo–Segura transfer, which will be discussed later, are a paradigm of large-scale investment based on the supply model giving rise to deep environmental, inter-regional and inter-sectoral unrest. Current attempts to reduce the unrest include the application of demand-oriented policies (based on water conservation and reuse) and supply-oriented measures (desalination plants) after plans for another transfer, this time from the River Ebro, were rejected.

With respect to the concessional regime and control over the hydraulic public domain, nowadays (since the Water Act of 1985) all waters (both surface- and groundwater) are public and subject to concessions.[2] The abstraction and use of new flows therefore requires the authorization of the corresponding basin authority (Confederación Hidrográfica) and is subject to modification upon change in the circumstances under which the concession was granted. But in practice, concessions are only reviewed in very exceptional cases. The basin authorities are responsible for the supervision of the hydraulic public domain and for enforcing compliance with concessions (both for surface- and ground-water) and also for closing illegal wells and exacting due sanctions.

Nevertheless, judging from the evidence, the means at the disposal of the basin authorities are, by any reckoning, insufficient, and, in our view, even structurally inadequate to enable them to undertake the type of control functions that are particularly necessary in order to guarantee efficiency under the demand-oriented model (nearly all civil servants in basin authorities are civil engineers specializing in hydraulic projects). As an illustrative example, at the end of 2001, the Segura basin authority estimated that some 10,000ha of land were being illegally irrigated (other studies place the figure between 25,000 and 30,000ha) and, in 1995, a major study undertaken by the Social Economic Council of the Region of Murcia estimated a total of 20,350 irrigation wells in the territory of the Segura basin authority, a figure that contrasts sharply with the mere 4500 that had been declared to the basin authority, 2574 of which had already been inscribed on its Register (public concessions) or Catalogue (private wells).

Furthermore, any decision to convert from dry to irrigated farming is subject to environmental impact assessment by the state authority. The fines imposed for partial or total failure to present the necessary documentation for a permit can be described as ridiculous in comparison with the potential profit at stake. At the same time, practice shows that the wells used for illegal irrigation are rarely closed and water trading goes on outside administrative control.

In short, the situation is one of relative impunity, since inadequate sanction and control mechanisms make illegal irrigation possible. Recall that, theoretically, a standard (amount granted) only meets the social optimum if there is an appropriate penalty for non-compliance.

Furthermore, as there is virtually no knowledge and control over the amount of water used for irrigation, there is no way of monitoring compliance with concession agreements, which are also quite outsized. The Water Law reform of 1999 made it compulsory to install either individual or collective flow meters and, although there are no precise data regarding the effectiveness of this ruling, the general impression is that it is still only minimally applied.

In such conditions, the difficulty of water consumption monitoring and thereby water conservation is almost impossible and is further aggravated by social and political pressures in order to find new resources to balance the deficit generated by new water demands. The former president of the Segura basin authority, Emilio Pérez, indicates that 'it is almost impossible to accomplish any improvement in water management in our country ... unless the current legislation is started to be applied on updating concessions and private uses. No management is feasible without knowledge on the reality to be managed.'[3]

Conflict among territories and dispersion of governance over matters relating to the water cycle

In recent decades, technology has made it possible to draw new resources into the water cycle, especially the reuse of water from filtering and desalination processes and from water-conservation systems based primarily on the modernization of existing irrigation systems (to reduce leakage in irrigation channels and drip irrigation systems). The bulk of the investment needed to access these new resources comes from the various tiers of governance (central, state and local) and to a lesser extent from private initiative.

There has been some serious confrontation between different administrations arising, in our opinion, from the lack of definition in the objectives of water policy and from clashing political interests (local or regional electoral motives blurring the longer-term perspective). In the state of Valencia, for example, every pre-election period sees a rekindling of the argument over the need to bring additional water resources from other regions and, in the case of the Júcar–Vinalopó transfer, also over the siting of the intake, which, incidentally, is based on the hypothetical savings that will result from the modernization of the Acequia Real del Júcar irrigation system. The state government also opposes central government plans for the construction of a desalination plant (in Torrevieja), while continuing to demand the reinstatement of the Ebro transfer project.

This same region, which has full autonomy in its territorial and agricultural affairs, has registered a sharp increase in the demand for water to serve new housing projects. Nevertheless, there is no coordination with the basin authorities, whose water resource availability reports for new housing projects are compulsory but not binding and therefore prove totally ineffective. Although both of the region's basins, the Júcar and the Segura, are characterized by structural deficit, in neither has it been possible to stabilize water demand. The problem of increased demand for non-agricultural purposes is aggravated by the above-mentioned lack of coordination in irrigation management. A high proportion of the new infrastructure, housing projects, etc. is taking over what used to be used for irrigated crop production. Water demand from the agricultural sector has therefore reduced but water concession agreements are rarely placed under review. Under such a framework, it is very questionable both in terms of efficiency and equality that cutbacks in periods of drought are the same percentage for all concessions, since some of them are extremely generous and others are quite scarce. In the state of Valencia the real amount per hectare can attain $60,000m^3/ha$ in some cases as a consequence of declines in irrigated acreage without any cutting down in concessions.

Furthermore, the extension of the irrigated area in the Albacete aquifer, which was promoted by the state government of Castilla–La Mancha, drastically reduced the flow of the River Júcar that fed the traditional irrigation system of Valencia. As a result, the central government has been forced to reduce the irrigation supply to the whole of the basin and even to purchase supplementary flow from Albacete during periods of drought. Supply is unlikely to balance with demand when each is in the hands of a different authority. In the absence of any coordination with well-defined objectives, decisions are often subject to short-term electoral interests. Another factor that may aggravate the total deficit in the basin is the practice of allocating new concessions (Júcar–Vinalopó transfer) based on hypothetical water savings, without checking and empirically quantifying the actual volume of water saved, as we shall see in the following section.

The hypothetical 'water markets', mentioned in the Water Law reform of 1999, have operated only on a testimonial basis, due to lack of surplus in the areas where it may be technically possible and the absence of adequate infrastructure with which to implement and supervise the water transfer.

Finally, in the past few years, amendments have been made to the autonomy statutes in several regions, clauses having been added in an attempt to protect the interests of individual regions. Thus, the statutes of Valencia and Murcia acknowledge the right to redistribute waters from basins in surplus (Article 20 of the state Statute of Valencia and the proposal of amendment to that of Murcia). The Statute of Aragon, which takes 'the opposite perspective, acknowledges the right to exercise vigilance to avoid the transfer of water from the hydrographic basins within its boundaries … for the sake of present and future generations' (Article 19). The Statute of Castilla–La Mancha makes direct reference to preferential rights over its own water resources to maximize availability to satisfy its own needs (Article 20). The Statute of Andalucía speaks of exclusive competence over the waters of the Guadalquivir basin flowing through its territory and not affecting other states (Article 51). That of Cataluña claims full autonomy over the water of its inland basins and participation in hydrological planning by the water management authorities of the central government (already included in federal legislation), while also accepting the compulsory duty to issue a regulatory report for any inter-basin transfer that may affect the level of the water resources within its boundaries (Article 117).

This brief summary of the 'logic' behind the so-called 'water wars' reflects not the ideologies of the country's major political parties, but electoral interests; hence the frequency with which the same political party can adopt different positions in each state. This is affecting not only the allocation of resources but also the range of policies being applied in matters relating to the water cycle.

The case of the Segura basin: The supply-oriented policies of the latter years of the Franco regime versus the mixed policies of today

A good example with which to verify the above and support our claims concerning matters of water consumption and water conservation is the Segura basin. This basin has seen in the space of a few years, both supply-oriented and demand-oriented initiatives, serious conflicts over water usage and simultaneous action by different public administrations affecting the water cycle.

The Segura basin spans an area of roughly 18,870km^2, distributed over four states: Murcia (60 per cent), Castilla–La Mancha (25 per cent), Andalucía (9 per cent) and Valencia (5 per cent). Its semi-arid climate is partly due to lack of rainfall (average annual rainfall around 400mm, two-thirds of which falls in the autumn) and partly due to its lithological composition (marl and saline clays). Evapotranspiration ranges between 600mm in the upper valleys and 950mm in the lower valleys and coastal strip of Murcia. The adversity of these atmospheric conditions is offset by the suitability of the climate for certain early crops, especially due to warm winters and long hours of sunshine.

The traditional irrigation system of the banks of the river Segura already served 65,000ha of irrigated farmland at the turn of the 20th century. Its subsequent evolution has been remarkable. A series of dams, the Alfonso XIII (built 1917), the Talave (1918), the Corcovado (1929) and the biggest of all, the Fuensanta (1932), with a capacity of 223hm^3, enabled the irrigation of 89,656ha of farmland by the year 1933.

The PAICAS (Integral Use Plan for the Segura Basin), approved in 1941, raised totally unreal expectations for the availability of water resources in the basin. The Decree of 25 April 1953 encouraged further demand for irrigation water, since, as well as setting up an order of preference for flow allocation, it permitted the legalization of all the pre-existing irrigated land, even some dating back less than the 20 years minimum stipulated in the Water Law of 1879. It also permitted extensions to the irrigation system in lands adjacent to the traditional

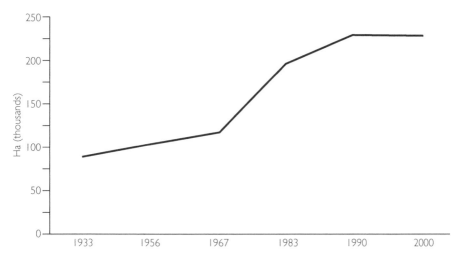

Source: Agricultural Yearbook, Ministry of Agriculture (2000).

Figure 11.1 *Evolution of irrigated land in Segura basin*

areas. In 1956 (before the inauguration of the Camarillas dam, the largest in the basin, with a capacity of 472hm^3) 104,420ha were already under irrigation.

A ministerial instruction was issued on 27 December 1966 determining that new irrigation areas (dated post-1953 and without consolidated rights) should be included in the allocation of resources, provided that this did not lead to the over-exploitation of available resources. In 1967 (the year in which the Tajo–Segura transfer obtained cabinet approval) there were already 117,230ha, of which only 63,917ha were permanently irrigated. Nevertheless, great expectations for the agricultural development of the area had already been built up and experience showed that illegal irrigation systems were eventually legalized. By 1983 (when, after a period of experimentation, a regular flow of water began), the irrigated area stretched to 196,874ha. Five years later it had reached the 230,000ha that remain to the present day (Figure 11.1).

The Tajo–Segura transfer set-up is a good example of the supply-oriented policy in force at the time. The objective was to serve a total of 269,000ha, comprising 147,000ha that were under-served irrigation-wise and 122,000ha of new irrigated land.

The draft project for the transfer was based on data relating exclusively to surface water and the deficit estimated to be covered by the transferred water allocated to the existing irrigated areas – an estimated amount of 380hm^3. Some authors (Sahuquillo-Herraiz, 1984; Senent-Alonso, 1984) have estimated that between 600 and 505hm^3 of groundwater was extracted in the early 1980s (before the water transfer) – just under half of that was renewable. In other words, the regulated resources plus the renewable groundwater left the existing irrigated area in a state of relative equilibrium.

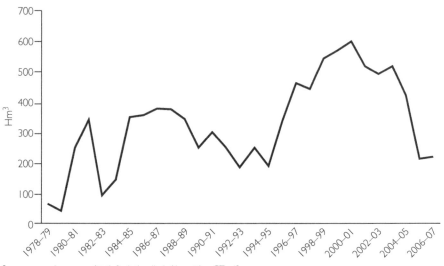

Source: www.chsegura.es/static/boletin_diario/AportAnuCT.pdf

Figure 11.2 *Annual volume of water transferred*

The spectacular expansion of the irrigated area that took place before the arrival of the transferred water was only possible at the cost of the overexploitation of aquifers and the withdrawal of river water, undertaken with a view to the citrus plantations being in full production by the time the transferred water arrived. The result of this was that, by the time the first transferred waters arrived, the River Segura was dead from its mid-course and had become a sewer tens of kilometres from its mouth.

The extended quote that follows is illustrative of the strategy employed by agricultural landowners: the engineer J. García, of the provincial delegation of IRYDA (the National Institute for Agricultural Reform and Development) in Alicante acknowledges in the *Plan for the Transformation of Irrigable Land in La Pedrera in Southern Alicante* (1981) 'that even in areas where the chance of water coming from the transfer is remote, there nevertheless persists the hope that the arrival of this water will reduce over-exploitation and thus help to restore aquifer levels … landowners far removed from the Pedrera reservoir expect their water prospecting results to improve with the water filtering from it… It is not known for certain which areas will be irrigated … and the prospect of obtaining water makes landowners reluctant to risk selling dry farmland only to see its value increase with the delivery of water from the transfer' (García, 1981 cited by Sanchis-Ibor, 2002).

This business strategy is not without its economic logic, since the private profit from irrigated land far exceeds that from dry land. By way of example, the study undertaken by the Region of Murcia Social and Economic Council (Consejo Económico y Social, 1995) using 1992 RECAN (Red Contable Agraria Nacional, or National Agrarian Accounting Network) data estimated gross added value per ha of irrigated land in the region to be 12 times higher than that of dry land.

Nevertheless, these expectations were not fulfilled. Of the 600hm^3/year predicted for the first phase (400hm^3 for irrigation purposes, 110hm^3 for public water supply and 90hm^3 in losses) the average amount transferred was 327hm^3 (just over half the amount predicted for the first phase and less than a third of the 1000hm^3 predicted for the second phase, on which the infrastructure was sized), 40 per cent of which went to public water supply, instead of the planned 18.3 per cent. In water shortage years, moreover, most of the water went to satisfy increased urban demand. Thus, from 1984 onwards, the proportion of transferred water allocated to irrigation averaged 40 per cent versus the 70 per cent originally planned.

Figure 11.2 shows the annual volume of water transferred. Note that in the drought years (1981–1984, 1991–1995 and 2005–2007), when the need was greatest, the volumes transferred are generally lower than in non-drought years, the only exception to this being the 1999–2000 hydrological year, when it practically reached the peak level.

The shortfall in volume transferred was made up in part with groundwater, leaving the aquifers extenuated, and in December 2006 the Ministry of the Environment[4] acknowledged that the situation in the south-east was one of illegal wells and an illegally run parallel water market. This was the outcome of an almost total lack of control by the basin authorities (leading to the 'tragedy of the commons'). It should be borne in mind, however, that the problem with the wells was due not so much to their illegality, although that did matter (drought wells and 1995 figures given earlier), as to the fact that in most cases landowners overstepped their rights not only with respect to the perimeter they had a right to irrigate, but also in terms of the volume of water allowed, the power of the pump and the depth of the well. In short, the relative equilibrium of the pre-transfer period was followed by an increase in water deficit.

One of the few positive factors worth mentioning, usual in supply-side policies, is that the transfer brought a substantial increase in income and wealth to the whole of the receiving region. It is also true to say that it gave a tremendous boost to farmland ownership concentration (Table 11.1) and an increase in the demand for salaried labour (Table 11.2) a large percentage of which was immigrant labour (both legal and illegal) which, while helping through low wages to sustain agricultural competitiveness, also requires major public initiatives to address problems of integration, supervision and social protection. Thus, once again, there is a stark contrast between private profit and social costs.

Table 11.1 *Ownership structure of irrigated farmland in Murcia (per cent)*

Farm size (ha)	1962	1972	1982	1989	1999
<10	54.4	45.7	37.7	39.7	27.5
10–49	23.8	24.3	29.1	27.4	26.8
50+	21.9	30.0	33.2	32.9	45.7

Source: Agricultural censuses of Spain (INE, 1962, 1972, 1982, 1989 and 1999).

Table 11.2 *Percentage of agricultural labour by state (1999)*

	Externalized	Family	Salaried
Andalucía	1.9	46.7	51.4
Aragon	1.8	72.4	25.7
Canarias	0.5	46.7	52.8
Cataluña	1.4	68.0	30.7
Valencia	17.2	55.6	27.2
Murcia	7.2	33.9	58.9
Spain	2.9	65.6	31.5

Source: Agricultural census of Spain (INE, 1999)

A wide variety of strategies have been used to deal with the water 'structural deficit'. Amendments to the National Hydrological Plan of 2001 in 2004[5] led to the cancellation of the Ebro transfer project, which allocated resources to offset the deficit and prevent aquifer overexploitation in the area without increasing irrigated acreage. Some groups continue to campaign for the Ebro water transfer to the present day, while the central government, through the public company ACUAMED, plans to increase supply by using desalinated water mainly for the public water supply. All groups are in favour of continuing with the Tajo–Segura transfer that has been challenged by the government of Castilla–La Mancha.

Demand-oriented policies consist largely of modernization plans for irrigation systems (introduction of drip irrigation) promoted by the administrations of both the central government, through ACUAMED (upstream waterworks) and SEIASA (downstream waterworks) and, to a lesser extent, by the state governments, and the reuse of filtered water to irrigate golf courses and, to a lesser extent, for agricultural irrigation, due to opposition from farmers who are particularly reluctant to use it because of water quality concerns and legal concessional problems.[6]

Regrettably, there are few reliable data on specific actions and the ways the investment has been funded.[7] The available information is patchy and, to our knowledge, at times inaccurate. Nevertheless, in subsequent sections, we will make use of data from other sources to discuss the achievements that have been made at least as far as water conservation initiatives are concerned.

In this case, therefore, the adoption of water-conservation technology was essential to sustain the irrigated area that a supply-oriented policy (the Tajo–Segura transfer) had encouraged, thereby generating serious conflict both with other basins (Tajo) and within the receiving basin (between farmers at the head and tail of the irrigation system) that have led to high social costs (pollution), attempted solutions for which have required public investment. The Region of Murcia is one of the states that has achieved one of the highest percentages of water savings in the whole of Spain, flood irrigation accounting for only 19.8 per cent of irrigated acreage in 2006 (Ministry of Agriculture, 2006).

The diversity of irrigated farmland in Spain

The first characteristic worth mentioning is the variety of edaphoclimatic conditions and the uneven distribution of water resources across Spain's territories. The southern states, Andalucía, Extremadura, Castilla–La Mancha, Murcia and Valencia, receiving the lower precipitations, concentrate 62 per cent of irrigation acreage (Ministry of Agriculture, 2006).

The Ministry of Environment figures for 2005 (Ministry of Environment, 2007) show that unit water consumption (m^3/ha) varied widely. The lowest consumption levels (less than 3000m^3/ha) are found in the inland basins of País-Vasco and Cataluña; levels between 3000 and 5000m^3/ha are found in the basins of Baleares, Galicia-Costa, Guadiana and Norte; between 5000 and 7000m^3/ha in the Guadalquivir, Duero, Segura, Júcar, Tajo and Mediterranea-Andaluza; and over 7000m^3/ha in the Ebro and Canarias basins.

Mean efficiency (water needs/plot consumption) also varies widely across states. Galicia, Asturias, Madrid, Aragon and Navarra achieve 60 per cent efficiency or less; País-Vasco, Valencia, Cantabria, Baleares, Castilla–La Mancha and Andalucía between 61 and 80 per cent efficiency, while Murcia and Canarias, where water resources are lowest, achieved more than 80 per cent efficiency.

According to the Ministry of Environment (2007), the ratio of net income between irrigated and dry farmland varies greatly across provinces and crop types. The ratio is lower in olive oil and wine production (the marginal return on water is lower) and, at provincial level, in Cordoba, Jaen and Malaga (olive producing regions) and in Ciudad Real, Rioja, Tarragona and Toledo (wine producing regions); the ratio is higher where agriculture is highly intensive (Alicante, Murcia and Almería) and in the very marginal dry farmlands (Zamora, Palencia, Burgos and Avila).

According to the agricultural census of 1999 (INE, 1999), the source of water for agricultural uses also varies widely across states (Table 11.3).[8] Nationwide, the irrigation acreage using surface waters accounts for 61.7 per cent of the total, while subsurface sources account for 37.1 per cent of acreage, and other sources (desalination and regeneration) for 1.2 per cent. By states, irrigated acreage with subsurface water is above the average in Baleares (91 per cent), Castilla–La Mancha (78 per cent), Canarias (75 per cent), Valencia (52 per cent), Murcia (46 per cent) and Andalucía (37 per cent). These are the island states and states in the south of Spain.

Table 11.4 gives the spatial pattern of irrigation concession systems, and is based on data from the agricultural census of 1999. Over Spain as a whole, two-thirds of the total irrigated acreage is managed by irrigation associations (almost all of them water user associations, WUAs), but the proportion increases to 85 per cent in Aragon, Navarra and Rioja. Water user associations have only a very minor presence in Baleares and Cantabria.

Table 11.3 *Percentage of irrigation acreage by source of water (1999)*[9]

	Subsurface	Surface	Other
Andalucía	37	61	2
Aragon	6	93	1
Asturias	11	84	5
Baleares	91	3	6
Canarias	75	15	10
Cantabria	6	77	17
Castilla–León	37	63	0
Castilla–La Mancha	78	22	0
Cataluña	23	75	2
Valencia	52	45	3
Extremadura	11	88	1
Galicia	16	83	1
Madrid	20	79	1
Murcia	45	52	3
Navarra	4	95	1
País-Vasco	9	77	14
Rioja	14	86	0
Total Spain	37	62	1

Source: Agricultural census of Spain (INE, 1999)

Table 11.4 *Distribution of irrigation by state and concession system (1999)*

	WUA ha	Individual ha	%WUA
Andalucía	522,900	309,600	63
Aragon	348,500	29,200	92
Asturias	2600	5600	32
Canarias	14,900	12,900	54
Castilla–León	244,600	194,300	56
Castilla–La Mancha	178,200	290,700	38
Cataluña	185,700	51,700	78
Valencia	231,100	52,500	82
Extremadura	192,900	43,000	82
Galicia	38,900	34,800	53
País-Vasco	8200	2400	77
Baleares	1500	16,300	8
Cantabria	100	1200	5
Madrid	10,800	15,000	42
Murcia	128,300	41,900	75
Navarra	67,200	4900	93
Rioja	28,300	4900	85
Total Spain	2,204,700	1,110,900	67

Source: Agricultural census of Spain (INE, 2003)

Table 11.5 *Distribution of irrigation by farm size and concession system (1999)*

Farm size (ha)	WUA ha	Individual ha	% WUA
<10	551,600	152,800	78
>10 – <50	749,200	289,800	72
>50 – <100	283,300	159,900	64
>100 – <500	434,700	319,000	58
>500	185,900	189,400	50

Source: Agricultural census (INE, 1999)

In general terms, water user associations are more common in traditional irriga-tion districts than in new ones and in districts using surface water than in those using groundwater. There is also positive correlation between farm size and water management system. As Table 11.5 shows, larger farms tend to have individual concessions, whereas smaller holdings tend to group together to manage irrigation water on a collective basis.

The water-conservation technology adoption process and its impact on water consumption

Current situation

In the National Irrigation Plan (NIP) 2000–2008, it is estimated that of the 3,770,000ha of irrigated land registered in 1996, 1,080,000ha are historic irrigated areas (pre-20th century), 320,000ha were created in the first third of the 20th century, 990,000ha are due to initiatives undertaken by the National Institute for Colonisation and IRYDA,[10] 100,000ha are due to initiatives of the states and 1,280,000ha to private initiatives with public grants (Ministry of Agriculture, 2001).

One-third of the irrigated area was served by earthen channels or poorly maintained lined channels. Two-fifths of the total surface area was flood irrigated and only one-sixth was drip irrigated. In one-third of the irrigated area, gross supply covered less than 75 per cent of water requirements. The NIP was aimed primarily at improving existing irrigation systems (reducing leakage and converting to water-saving application techniques), rather than undertaking large-scale conversions of dry land into irrigated areas, and at creating new irrigated areas only when they were deemed to be of social interest (small-scale irrigation systems in less favoured rural areas) and concluding projects in progress.

Funding for these actions was undertaken by public administrations (the Ministry of Agriculture and the states) and by private initiatives (50 per cent on programmes for the consolidation and improvement of irrigated areas and around 23 per cent on irrigation projects in progress and of social interest).

Table 11.6 *Distribution of irrigated area by states (2006)*

	Flood		Sprinkler		Self-propelled		Drip		Others and missing data	Total	
	ha	%	ha	%	ha	%	ha	%	ha	ha	%
Andalucía	186,939	20.15	94,309	10.16	24,107	2.60	615,619	66.34	6984	927,958	99
Aragon	234,424	61.87	67,511	17.82	36,788	9.71	39,526	10.43	620	378,869	100
Asturias	1205	74.34	127	7.83		0.00	214	13.20	75	1621	95
Baleares	2196	13.56	3622	22.37	1698	10.49	8523	52.65	150	16,189	99
Castilla–León	183,730	44.31	128,581	31.01	83,005	20.02	14,006	3.38	5336	414,658	99
Castilla–La Mancha	28,642	6.19	109,199	23.60	81,065	17.52	240,587	52.00	3180	462,673	99
Canarias	1951	8.81	3433	15.51		0.00	16,479	74.44	274	22,137	99
Cantabria		0.00	208	64.00		0.00	117	36.00		325	100
Cataluña	146,562	58.70	21,637	8.67	8409	3.37	71,438	28.61	1619	249,665	99
Extrema-dura	107,455	53.11	28,071	13.87	13,779	6.81	52,393	25.90	618	202,316	100
Galicia	22,797	78.40	5021	17.27	336	1.16	728	2.50	194	29,076	99
Rioja	12,499	38.51	6027	18.57		0.00	9305	28.67	4625	32,456	86
Madrid	13,650	67.86	4069	20.23	892	4.43	1316	6.54	187	20,114	99
Murcia	33,080	19.77	1601	0.96	15	0.01	132,073	78.93	565	167,334	100
Navarra	59,191	67.08	11,873	13.45	429	0.49	15,518	17.59	1232	88,243	99
País-Vasco	607	6.97	6274	72.02		0.00	1212	13.91	618	8711	93
Valencia	128,325	43.14	1866	0.63	1076	0.36	162,781	54.73	3396	297,444	99
TOTAL	**1,163,254**	**35.04**	**493,430**	**14.86**	**251,598**	**7.58**	**1,381,835**	**41.62**	**29,674**	**3,319,791**	**99**

Source: ESYRCE by Ministry of Agriculture (2006)

Although the results of the Crash Plan to Modernize Irrigation approved by Royal Decree 287/2006 and have yet to be evaluated, the available data show that most of the objectives had been met by 2006.

According to the ESYRCE (Ecuesta sobre Superficies y Rendimientos de Cultivos) survey on crop acreage and yields (Ministry of Agriculture, 2006), in 2006, a total of 3,319,800ha of Spanish farmland was under irrigation – 35 per cent by flooding, 42 per cent by drip and 15 per cent by sprinkler (the remaining 8 per cent consist of other forms of irrigation or missing data). When it comes to the spatial pattern of irrigation methods (Table 11.6), flood irrigation predominates in Galicia, Asturias, Madrid, Navarra and Aragon, while drip irrigation is the main system in Murcia, Canarias, Andalucía, Valencia, Baleares and Castilla–La Mancha, all of which are characterized by shortage of water relative to demand.

According to the ESYRCE survey (Ministry of Agriculture, 2006) by broad crop category (Table 11.7), the main method for grain, forage and extensive herbaceous crops is flood irrigation, whereas drip irrigation is the main technique used with intensive woody crops – citrus and fruit trees – and for irrigated olive trees and grape production.

According to the Ministry of Agriculture (Figure 11.3), which provides data only from 2002 onwards, the irrigated acreage has increased slightly from 3,312,000 to 3,318,800ha. It is worth noting the substantial increase in drip irrigation, mainly to replace the flood system.

Table 11.7 *Distribution of irrigated acreage by crop and irrigation system (2006)*

Crop group	Total acreage	Flood		Sprinkler		Sprinkle automotive		Drip	
	ha	ha	%	ha	%	ha	%	ha	%
Cereals	886,896	520,712	58.7	206,044	23.2	138,411	15.6	15,829	1.8
Citrus	307,173	87,761	28.6	220	0.1		0.0	217,878	70.9
Forages	281,524	167,028	59.3	69,860	24.8	40,540	14.4	446	0.2
Fruits	251,784	82,257	32.7	3840	1.5	687	0.3	164,267	65.2
Vegetables	205,967	41,767	20.3	41,389	20.1	14,920	7.2	107,042	52.0
Industrial crops	201,274	60,423	30.0	94,823	47.1	29,699	14.8	14,818	7.4
Grain pulses	21,895	5179	23.7	9826	44.9	5219	23.8	1582	7.2
Olive trees	555,673	40,116	7.2	2597	0.5	2458	0.4	506,262	91.1
Tuberous crops	46,618	8624	18.5	29,926	64.2	4423	9.5	3342	7.2
Vineyard	317,637	12,561	4.0	15,376	4.8	3430	1.1	283,315	89.2

Source: ESYRCE by Ministry of Agriculture (2006)

Table 11.8 shows the variations in the areas irrigated by flood and drip irrigation systems by state, and the variation in the percentage distribution of irrigation systems across the total irrigated area between 2002 and 2006. There is a reduction in the flood-irrigated area in all regions, except for Asturias and Baleares where there has been a slight increase, mainly due to a switch to the drip method.

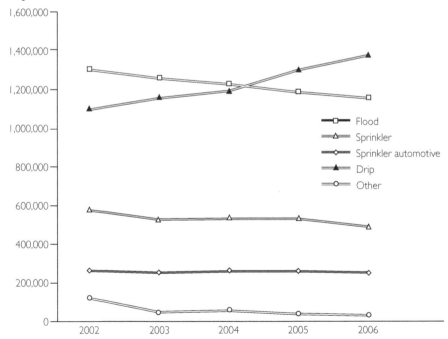

Source: ESYRCE by Ministry of Agriculture (2006)

Figure 11.3 *Evolution of irrigated acreage by irrigation system*

Table 11.8 *Variations in irrigated acreage by state for the period 2002–2006*

	Flood	Drip	Total	Variation in drip acreage
	ha	ha	ha	%
Andalucía	−27,302	127,800	84,279	9
Aragon	−44,324	5195	−32,168	2
Asturias	1137	−131	1208	−70
Baleares	1369	2093	2022	7
Castilla–León	−49,562	6924	−49,823	2
Castilla–La Mancha	−13,027	69,592	15,221	14
Canarias	−5264	5438	−1783	28
Cantabria	−392	117	−327	36
Cataluña	−11,867	6569	−4780	3
Extremadura	−19,837	21,751	−11,375	12
Galicia	−1253	−50	−637	0
Rioja	−4954	−193	−8615	6
Madrid	−1851	792	−762	4
Murcia	−6439	6011	−3952	5
Navarra	−14,308	15,472	1870	18
País-Vasco	−435	667	−184	8
Valencia	−18,099	60,330	−21,538	23
Spain	−234,918	334,257	−34,381	10

Source: ESYRCE by Ministry of Agriculture (2002, 2006)

The area of land under drip irrigation has increased, especially in Andalucía, Castilla–La Mancha and Valencia in that order. The biggest increases in the ratio of drip irrigation to total irrigated area (an indicator of the effort to modernize irrigated farming in recent years) in regions that were already drip irrigating on a large scale are found in Canarias, Valencia, Castilla–La Mancha and Andalucía in that order.

Estimation of changes in agricultural water consumption in Spain

In this section we estimate variations in plot water consumption in Spanish agriculture between the early 1990s (the average for 1992, 1993 and 1994) and the first years of the 21st century (the average for 2003, 2004 and 2005) by state and the broad crop type.

The procedure uses Ministry of Agriculture data (Agricultural Yearbook by Ministry of Agriculture, 1992, 1994, 2003 and 2005) to calculate the area dedicated to each crop category, calculate the difference in area between the two periods and assign to each crop its average unit water consumption.[11]

By using average consumption per crop, we ignore differences in consumption due to edaphoclimatic conditions or the irrigation method used (the most widespread method in the last period considered was drip irrigation). As a result, the figures shown are only an approximation to the real values and should there-

Table 11.9 *Changes in the irrigated acreage of herbaceous crops (1992–1994 versus 2003–2005, ha)*

	Vegetables		Potatoes	Grain and pulses	Flowers and ornamental plants		Industrial crops	Forages	Cereals	Rice	Corn
	Open air	Protected			Open air	Protected					
Galicia	436	1754	−3585	−4843	273	289	−11	−11,584	−18,074	0	−17,975
Asturias	165	23	0	27	0	1	0	127	0		0
Cantabria	−24	−15	26	−2	2	4	3	−139	−3		−3
País–Vasco	221	55	321	5	19	17	306	191	14		14
Navarra	2967	95	−469	2741	10	8	−8079	2474	4215	632	1438
Rioja	−3429	−4	−3639	−375	12	2	−1582	−1657	2344		314
Aragon	−5499	−60	−4702	4927	−68	−68	−48,903	25,610	39,390	3423	41,296
Cataluña	−7884	−351	−4529	180	130	277	−6582	−4875	13,826	33	14,514
Baleares	1565	215	−1096	67	−8	−35	−86	−4782	836	49	90
Castilla– León	−3955	−132	−432	7841	−2	3	−27,983	−27,273	139,476	211	77,208
Madrid	−984	−1200	−1159	3580	−8	47	−1609	−101	1025		860
Castilla– La Mancha	−3632	−1467	−6300	7880	−476	−193	−71,314	−7811	82,008		4420
Valencia	−9989	−1092	−4306	−72	923	257	−2245	−2755	203	−724	−1416
Murcia	9052	1903	−1332	−45	127	204	−3118	−2090	−2454	166	−1143
Extrem- adura	1114	1613	−4872	−1376	−155	−81	−64,945	2233	40,353	15,388	38,658
Andalucía	1100	13,394	−6417	14,453	−61	884	−73,293	−10,445	59,806	29,685	27,887
Canarias	−1173	−862	−1714	−21	−605	−543	−7	−107	−244		−235
Total Spain	−19,949	13,870	−44,206	34,934	1729	2350	−309,448	−45,277	367,399	53,284	185,929

Source: Agricultural Yearbook, Ministry of Agriculture (1992, 1994, 2003 and 2005).

fore be treated with caution. Nevertheless, we consider them a close enough approximation for our purposes. Furthermore, these are moderate estimations since the states in which water consumption has increased most are those of the south, with consumption above the mean used.

The irrigated acreage in Spain has increased between the two periods considered by 11 per cent to 385,000ha (of which 384,000ha were for woody crops and 1000ha for herbaceous crops). However, there were significant changes in crop composition and, as a result, an increase in water consumption for irrigation purposes. The data therefore reflect, with the limited accuracy provided by the administration, changes in the amount of land devoted to each crop and give some indication of the variation across regions and crop groups.

The changes in the acreage of herbaceous crops during the period by broad crop group are shown in Table 11.9. Although the increase in the irrigated acreage of herbaceous crops has been only 1400ha, there are some significant increases: vegetables in Murcia, grain and pulses in Andalucía, Castilla–León, Castilla–La Mancha and Aragon, rice in Andalucía and Extremadura and, above

Table 11.10 *Changes in the irrigated acreage of woody crops (1992–1994 versus 2003–2005, ha)*

	Citrus trees	Pip fruit trees	Stone fruit trees	Olive trees	Wine grapes	Table grapes	Dried fruits	Other woody crops
Galicia	58	528	439		−33		49	−288
Asturias	0	0	0				0	54
Cantabria	−10	−8	−5				0	10
País-Vasco	0	−17	0		2625		0	9
Navarra	0	115	−421	1131	8008		384	141
Rioja	0	−841	−1560	941	3243		160	5
Aragon	0	−4291	4676	3700	5745	21	2202	−29
Cataluña	4168	−5067	396	8244	1704	2	−2791	52
Baleares	228	−377	35	114		−18	404	267
Castilla–León	0	−1347	−117	231	1220		−22	40
Madrid	0	−257	−63	393	166		7	0
Castilla–La Mancha	0	−948	−476	12,275	87,760	−23	1884	39
Valencia	894	−1731	−5515	5724	4387	−3315	−7692	1213
Murcia	5678	−348	−2957	4880	2381	1540	−7091	−279
Extremadura	−21	−2054	6490	4193	1500	−2	135	−23
Andalucía	28,191	−1418	2887	209,075	43	−2164	2281	406
Canarias	−178	31	−98	10	471	−20	4	1347
Total Spain	38,999	−17,703	3637	250,921	119,229	−4014	−10,088	2961

Source: Agricultural Yearbook, Ministry of Agriculture (1992, 1994, 2003 and 2005)

all, corn in Castilla–León, Aragon, Extremadura and Andalucía. These last two crops, rice and corn, are the most water consuming.

The changes in the acreage of woody crop groups are shown in Table 11.10. The larger increases in acreage are for citric trees in Andalucía, Murcia and Cataluña, stone fruit trees in Extremadura, Aragon and Andalucía and, above all but with low water requirements, olive trees in Andalucía and Castilla–La Mancha, and vineyards in Castilla–La Mancha.

Table 11.11 shows variations in the total volume of water used for irrigation purposes in each state, by crop group broadly categorized into woody and herbaceous. A positive value represents an increase in volume and a negative value a decrease. Consumption across the whole of Spain increased by almost 1500hm³, roughly 7 per cent of the water consumed in the past few years. The increase was highest in Andalucía (woody and herbaceous crops), Aragon (mainly herbaceous), Castilla–León (herbaceous) and Extremadura (mainly herbaceous). Water consumption decreased in some regions, in particular Valencia and Galicia.

By crop category (Table 11.12), the highest increase in consumption can be observed in cereals, mainly due to the increase in the acreage used to cultivate maize and rice; the next highest increase is in olives, citrus (due to geographical expansion) and vineyards (at the beginning of the study period, the irrigation of

Table 11.11 *Change in irrigation water use by state (1992–1994 versus 2003–2005, hm³)*

State	Woody	Herbaceous	Total
Galicia	4.1	−129.6	−125.5
Asturias	0.2	1.0	1.2
Cantabria	−0.1	−2.0	−2.1
País-Vasco	3.9	5.6	9.6
Navarra	12.9	33.5	46.4
Rioja	−4.3	−40.3	−44.6
Aragon	19.2	309.2	328.4
Cataluña	12.7	−27.0	−14.3
Baleares	0.9	−26.0	−25.0
Castilla–León	−4.2	294.3	290.1
Madrid	−0.6	−6.7	−7.3
Castilla–La Mancha	144.5	−120.8	23.7
Valencia	−39.6	−95.5	−135.1
Murcia	22.6	9.3	32.0
Extremadura	28.5	169.2	197.7
Andalucía	477.4	382.9	860.2
Canarias	3.1	−22.8	−19.7
Total Spain	682.1	786.2	1468.3

Source: Agricultural Yearbook, Ministry of Agriculture (1992, 1994, 2003 and 2005)

olive trees and vineyards was merely symbolic). There is a decline in water consumption associated with forages, potatoes and pip fruits.

As the increases in water consumption at plot level are close to the increases in irrigated acreage (10 and 11 per cent, respectively), it seems that the volume of

Table 11.12 *Change in irrigation water use by crop (1992–1994 versus 2003–2005, hm³)*

Crop	
Citrus	214.5
Pip fruits	−79.7
Stone fruits	16.4
Other fruits	8.9
Dry fruits	−15.1
Olive trees	376.4
Vineyards	160.8
Vegetables	−24.3
Potatoes	−176.8
Grain and pulses	69.9
Flowers and ornamental plants	19.5
Industrial crops	−737.0
Forages	−216.6
Cereals	1851.5
Total	1468.3

Source: Agricultural Yearbook, Ministry of Agriculture (1992, 1994, 2003 and 2005)

water per hectare has not changed much. However, the investments in modernization (improvement of conveying channels and adoption of saving technologies at plot level) imply water savings that are difficult to estimate. Assuming that the average savings are at least 20 per cent if adopting drip irrigation, total savings will be more than 500hm^3 per year – these savings have been used to cover the requirements of increasing the acreage of the more water-demanding crops. Therefore, the savings of water from advanced irrigation technologies do not translate into total water savings for the Spanish irrigation system, and potential savings from efficiency gains driven by investments do not materialize.

Motives for the adoption of water-conservation technologies in Spanish irrigation

We have already shown how major progress is being made in the use of irrigation water-saving technologies in Spanish agriculture, driven by the growing recourse to aquifers and also because of the escalating water scarcity. In our view, due to the wide range of conditions across Spanish irrigated agriculture, the motives that may sway the decision of a farmer or his irrigation association to adopt this type of technology are varied and complex, and not in all cases related to water conservation, despite that being the main objective cited by public administrations in their plans for the modernization of irrigation. Nevertheless, public administrations have made a noteworthy effort in this respect over the past decade and any progress is largely due to them.

The central government recently approved the above-mentioned Crash Plan for Irrigation Modernization which aims, by 2007, to have modernized 866,900ha with a total investment of €2344 million (80 per cent from the federal Ministries of Agriculture and Environment and the rest from users).[12] Although the data available at the time of writing are insufficient to properly assess the plan, it is worth noting that the initiative is strictly that of the central government with no participation by states.

The Crash Plan therefore continues the Spanish tradition whereby water management issues are left to the control of public administrations. While the objective in the past was to deliver water to users, it is now to save water by modernizing the irrigation system and introducing new technology; the bulk of the funding, in all cases, comes from the public administrations.

An innovation in recent decades is for the investment to be undertaken by a series of institutions and participating bodies: users (mainly water user associations), federal ministries, federal public companies (SEIASAS and TRAGSA both dependent on the Ministry of Agriculture; water companies dependent on the Ministry of Environment) and state ministries.

A case in point, where all the above-mentioned entities are participating, each independently and in its own right, is the modernization of irrigation in the Acequia Real del Júcar (a water user association established in 1273 and responsible for the irrigation of over 20,000ha). This has naturally given rise to coordination problems and misalignments between the local boards distributing the flow in different municipalities.

The Albufera Natural Park in Valencia is a lake located between the Acequia Real del Júcar and the sea, and the irrigated area closest to the lake is used to cultivate rice. As a result, traditionally, the flood irrigation practices that were pervasive until recent times meant that water was bound to enter the reservoir in the form of irrigation returns. From now on, water that used to enter the Albufera of Valencia will have to be made up with water reclaimed from urban treatment plants, especially from the city of Valencia, with all the investment that that involves and with uncertain quality standards.

The low cost of water for irrigation purposes, its abundance and the prevalence of smallholdings in the area were the factors taken into account by farmers deciding whether or not to maintain flood irrigation. The modernization process that has been undertaken includes pressurized secondary channels, improvements in primary channels, and pressurized distribution networks and irrigation heads (up to the irrigation area or plot, as may be decided) which, in the future, will enable every farmer to install drip irrigation. The water saved in this way will, as already mentioned, facilitate the Júcar-Vinalopó transfer, although the Acequia Real claims that the abstraction point for this transfer should be made close to the mouth of the river to ensure water availability.

Although, as we have seen, public administrations in Spain play a key role in decision making, farmers also can decide to make the necessary investment to make the changeover from flood irrigation to other techniques, very often with the aid of public subsidies. There are widely varying reasons to induce farmers to install water-saving technology; they will depend in each case on the expected return to the investment.

Studies conducted by the Centro Valenciano de Estudios de Riego (CVER) have analysed the behaviour of farmers in Valencia, using data collected at interviews with the irrigators and managers of irrigation bodies. During the 1990s, the motives behind the decision to adopt this type of technology were to save on labour and improve product quality (García-Mollá, 2000). According to irrigating farmers in Valencia, it is not worth making the necessary investment for the water savings alone, considering water availability and prices. This fieldwork also revealed that, generally speaking, irrigation entities using groundwater tended to keep their installations in better condition. This is because the more water they are able to save, the further they can extend the irrigable area. Irrigation entities using surface water, on the other hand, are unable to extend in this way. In cases such as these, it is usually the public administration that takes the initiative to improve the irrigation systems in certain areas.

Apart from the availability of groundwater in some areas (Albacete and examples given above), other factors that induce farmers (or irrigation associations) to take up more efficient irrigation techniques are water shortage (in Murcia and Almería, for example), the design for quality improvements and the reduction of labour needs (especially in the cultivation of woody crops) and the need to adapt to the chosen technological package (examples being strawberry cultivation in Huelva with raised gutters or padding technologies, and greenhouse cultivation technologies).

Consequences of the adoption of water-conservation technology

The supply-oriented water management model has predominated in Spanish history and continues to have a strong influence on current policy. The demand-oriented model which opts for the use of techniques to rationalize the demand for water as a scarce factor of production is beginning to have some impact in Spain, but to what extent has it led to significant water savings? It is fairly obvious that improvements in irrigation networks must lead to some degree of reduction in water demand, but it is less obvious when we focus on the basin or demand unit.

Studies conducted by the CVER in the state of Valencia, in the late 1990s, showed that irrigation associations using groundwater tended to use the water more efficiently, by installing water-saving technology, keeping irrigation equipment maintained and, in some cases, by enforcing the rule that plots should be properly prepared before irrigation (Carles-Genovés et al, 1998; García-Mollá, 2000). This research concluded that the nature of the water right largely determined the technological infrastructure that would be used, since entities using groundwater extend their networks to reach additional plots as farmers buy more shares in the well. The only restriction imposed by practically all of these companies at the time of their constitution is the maximum area allowed to be irrigated. The precise area that is or may be irrigated is determined *a posteriori*, according to the location of the plots owned by the farmers who purchase 'shares' in the well or wells belonging to the company, and the irrigation network is subsequently extended to reach each and every one of the plots owned by those who have obtained the right.

If, in addition to this peculiar network design, we also consider the very small average size of these companies, we will find many municipalities in which the networks of different companies cross at numerous points, and plots located in close proximity to a well are being irrigated from another situated several kilometres away; or adjacent or very close plots being irrigated from different wells far removed from either, due to their owners having rights from different companies. Installations intended to improve technical efficiency will therefore only be as efficient in saving water as the irrationality of the irrigation networks will allow. There are also cases in which water savings lead to the issue of new shares and thence to further conversion of dry to irrigated farmland. Thus, although some saving of water per unit area is achieved, enlargement of the area leaves total water consumption unchanged.

The Vall d'Uixó, in La Plana Baixa county of the province of Castellón, is a case in point. During the 1990s, all of the water used for irrigation purposes within the municipal boundaries was managed by collective entities, all of them companies dating prior to the current water law, and therefore the exploitation regime of groundwater was private.

Irrigation rights were acquired through the purchase of shares, some of which were linked to land (cooperatives) whereas others were not. These shares confer the right to irrigate a particular area. As local farmers converted their plots to irrigation and acquired all the necessary shares from one company or another, the area to be irrigated by each entity was gradually defined. In this way, the infrastructure needed to reach the plots owned by the shareholders was progressively installed. This process resulted in the creation of a crisscrossing irrigation network and duplication of pipelines, frequently resulting in efficiency losses and no less often in impediments to rational resource management. The situation eventually became unsustainable, since the aquifer from which most of the resource was being abstracted was both overexploited and salinized, while the irrationality of the irrigation system was driving up water consumption per unit area to very high volumes.

The scarcity and progressive salinization of water resources due to over-exploitation of the aquifer led the Júcar basin authority to order the establishment of a general water user association of the Vall d'Uixó,[13] with a view to regulating groundwater use within the municipal boundaries. The association is formed by the municipal council – to cover urban consumption – and by every one of the 16 well companies and two irrigation cooperatives operating within the municipal boundaries. At the time, the association operated only in an administrative capacity and the private entities maintained their managerial independence.

The total area under the management of the irrigation entities that made up the general water use association of the Vall d'Uixó amounted to 2822ha, theoretically the entire irrigated area within the municipal boundaries. The Júcar basin authority had banned any further abstraction of water and hence any extension of the irrigated area during the mid-1980s. The municipal boundary had a total of 40 wells and a concession of urban wastewater.

All companies used a similar organizational approach. The water was distributed by the companies or cooperatives, each to its own members through its own piping system. Farmers had access to the water on demand; there was no turn taking. Whenever some members of the board of the general water user association suggested forcing the companies to introduce irrigation on demand by turn, it was rejected by majority vote. It was the irrigation manager of each association who controlled the flow and the hours of service. Overwhelmingly, the most widely used irrigation technique was furrow flooding. Although drip irrigation was beginning to spread on large plots by individual initiative, and on a collective basis by some companies, this was only on a minor scale.

This situation motivated irrigation entities to join together to try to improve management. Nowadays, practically all the irrigation entities in the municipality participate in the joint management of resources and have changed over to drip irrigation – except for approximately 80ha – cutting out work for irrigation managers. The farmers of the area have estimated that management improvements and the introduction of drip irrigation have led to a 42 per cent reduction in water consumption compared with the previous system.

This is an example of how the use of water-saving techniques, together with water-management improvements, can help to improve the state of the aquifers. Nevertheless, there are also many examples of areas where water-saving techniques have been adopted on a widespread basis, resulting in highly efficient water application rates, but also in the overexploitation of aquifers. Large amounts have also been invested by public administrations and private agents, but, as shown in the section entitled 'Estimation of changes in agricultural water consumption in Spain', this does not necessarily imply water savings. While the amount of water saved is not used to increase reserves and the irrigated area continues to be extended, global demand will not fall. Therefore, demand-oriented policies will not have the desired effect without the introduction of restrictions on water abstractions and the size of the irrigated areas.

Conclusions and policy implications

The political organization of territories in Spain, and the distribution of competencies among the different administrations (federal, states and municipalities), requires coordination among the political and social agents in matters of water policy. The absence of coordination has triggered important conflicts derived to a large extent from the need to design and execute policies for the medium and long term, rather than merely responding to short-term electoral interests. Each major political party has maintained a different standpoint in water issues, beyond political ideologies. Sometimes these 'water wars' brought about large effects on the implementation of policies for water conservation, resulting in increased demand. Social and political criteria should be clearly defined as well, in order to decide on the appropriate measures for solving contradictory objectives, such as recovery of overdrafted aquifers and continuation of existing irrigation.

Additionally, land use planning and agricultural or other related policies, should be linked with the water policy, defining clearly the objectives and constraints of each policy. For example, water availability is disregarded as a constraint that should be fulfilled in land use policies, and also the promotion of new irrigation projects is difficult to understand without due consideration to the inner meaning of the 'water wars'.

The long-established and prevailing role of the administration in the allocation of resources and the execution through privileged financing of hydraulic works has generated an important resistance to changes. These changes are required to move from a supply model towards a demand model, and also for the implementation of the European Water Framework Directive. This directive pursues the environmental sustainability, which sometimes could be in conflict with the economic and social sustainability. The inertia from the past in water management occurs both in public administrations and among stakeholders and large companies in charge of public works, since all of them are unwilling to reorient their strategies.

The lack of effective control mechanisms over concessions is an important management failure. It could imply that when the water supply is increased in order to balance a deficit in some territory, the result could be an even larger increase in demand which enlarges the water deficit. To avoid these situations, it is essential to improve the knowledge of existing irrigation districts, their concessions and their water requirements, and also apply all the rules and regulations from binding legislation.

Knowledge of irrigation in Spain has improved dramatically over the past few years. These improvements include estimates of irrigated acreage, quantity and quality of water bodies, and results from the various economic studies undertaken. However, some statistical sources need further improvement, such as irrigation acreage (where large discrepancies among sources remain). The Instituto Nacional de Estadística (INE) survey carried out by mail needs to be enhanced with 'in situ' verifications, checking the real water types of use, the user characteristics and the volumes really spent. In this regard, the unqualified usage of some of these statistical sources could produce quite absurd outcomes. Because of the large diversity of irrigation in Spain, the policies have to be differentiated for a correct implementation in the various territories. Therefore, the required studies have to be improved at the regional level.

There has been an impressive advance in water savings in Spanish irrigation during the past decade, through the enhancement of channel networks and parcel irrigation technologies, especially in regions with substantial water scarcity. The public initiatives have been crucial for the modernization of Spanish irrigation.

An overall approximation to the evolution of water consumption in the past decade is possible because of the available homogeneous data on irrigated acreage. These data indicate that although the increase in irrigated acreage (11 per cent) is similar to the increase in water consumption (9 per cent), the increase in acreage is linked to high-water-demanding crops, and therefore the important water savings resulting from irrigation modernization have made it possible to stabilize total water consumption. Also, important changes have occurred in the distribution of water among states.

The process of irrigation modernization over the past few years has been driven by public administrations, although sometimes there have been coordination problems among them. One problem is the lack of firm knowledge on both the water-saving potential from projects, and the possibility and costs of expanding the supply from new resources (mainly desalination). This knowledge requires making comprehensive economic, social and environmental studies that guarantee water conservation and the effective use of new resources. The studies should provide details on user characteristics, water pricing, water extractions, evapotranspiration and returns to watersheds, pollution loads and impacts on aquatic ecosystems.

The motivations for adopting water-saving technologies are diverse. The common objective of public administrations is to achieve water savings and ensure supply in areas with severe scarcity. The larger farms, based either on surface or subsurface water, adopt saving technologies for economic reasons derived from economies of scale that reduce production costs, and also to improve operations and product quality. The farms based on subsurface water adopt technologies to expand acreage or to reduce aquifer overdraft in order to lessen salinity problems. In contrast, small or medium-sized farms (which are very parcelled) with flood irrigation systems, only adopt from the 'pressure' of public administrations provided that the supported costs are low. The current water price is not an important factor in adopting water-saving technologies, at least in traditional Mediterranean irrigation. The scarcity of the resource, rather than price, has been the main factor explaining adoption of advanced technologies.

In order for the water savings in plots and channels to translate into savings for whole watersheds or districts, the changes in irrigated acreage and in the water needs of the crop mix have to be taken into account. Sometimes, the achieved savings are made available to expand the irrigated acreage and total water consumption remains constant, especially in the case of subsurface water. In other cases, the crops planted after technological adoption are more water-demanding than previous crops.

Notes

1   Article 148. The states can assume the following functions: regional/spatial planning, land use planning and housing; public works of regional interest; agriculture and livestock raising in accordance with general economic policy; building and operating hydraulic infrastructures and resources of regional interest (e.g. channels and irrigation projects); the promotion and planning of tourism within their boundaries.
Article 149. Functions that are considered to be of the exclusive competence of the central government: setting the basis for and coordinating the general planning of economic activity; legislation and regulation of water resources when they flow over more than one region.
2   Except for aquifer waters, where owners could decide to maintain their private ownership, which is under some restrictions. Obviously, a majority of owners kept their rights.
3   See Pérez-Pérez (1997), p353.
4   Ministry of Environment (2006), p29.
5   Royal Decree-Law 2/2004, amending the National Hydrological Plan Law 2/2001.
6   New regulations on treated water reutilization and regeneration have been recently issued by the Royal Decree-Law 1620/2007.
7   Clear data regarding the budget are available but not regarding the actual investment, and there are only global data regarding the source of the funding by public companies or administrations.

8   Although official statistics on irrigation in Spain have improved substantially, and those on irrigation acreage and crops are quite good, other important variables such as source of water, volume provided, primary and secondary channel losses, and plot irrigation systems are not so good. The two sources in need of substantial improvement are the survey on water use to irrigation user associations, collected by mail, and the water accounts, both from Instituto Nacional de Estadística (INE).

9   The survey on irrigation water from the INE also provides these data, by collecting information from water user associations and other entities by mail. The problem is that it excludes all individual water concessions, and although the INE tried to estimate these after 2005, the results are quite poor. For example, the survey by the INE estimates the irrigation acreage with surface water in Valencia at 81 per cent of total irrigated acreage. Our studies indicate that approximately one-third of acreage is irrigated with surface water, one-third with subsurface water and the other third with mixed water. Other sources such as the agricultural census or the National Irrigation Plan state that in Valencia, the acreage irrigated with subsurface water is above the acreage with surface water.

10   National Institute for Agricultural Reform and Development.

11   The volumes considered (m³/ha) were as follows: vegetables, 4000; flowers and ornamental plants, between 4500 and 5000; grain and pulses, 2000; industrial crops, between 1500 and 4000 for sugarcane and beetroot; forage, between 1500 and 7500 for alfalfa; cereals, between 1500 and 12,000 for rice; citrus, 5500; pip and stone fruits, 4500; olive trees, wine grape and dry fruits, 1500; other woody crops, between 2000 and 5500.

12   The states may participate if they reach an agreement with the central government to co-fund the investment.

13   'General' means that it is made up of base level entities, such as water user associations.

References

Carles-Genovés, J. (2000) 'La Administración Pública ante las nuevas políticas de aguas de la Directiva Marco', Paper presented at the II Congreso Ibérico sobre Planificación y Gestión de Aguas, Oporto

Carles-Genovés, J., Avellà-Reus, L. and García-Mollá, M. (1998) 'Precios, costos y usos del agua en el regadío mediterráneo', Paper presented at the Congreso Ibérico sobre Gestión y Planificación de Aguas, Zaragoza

Consejo Económico y Social (1995) *Estudio Sobre los Recursos Hídricos y su Importancia en el Desarrollo en la Región de Murci,* CES Región de Murcia, Murcia

García-Mollá, M. (2000) 'Análisis de la influencia de los costes en el consumo de agua en la agricultura valenciana. Caracterización de las entidades asociativas para riego', PhD thesis, Centro Valenciano de Estudios sobre el Riego, Universidad Politécnica de Valencia, Valencia

INE (1962, 1972, 1982, 1989 and 1999) *Censo Agrario*, Instituto Nacional de Estadística, Madrid

Ministry of Agriculture (1992, 1994, 2003 and 2005) *Anuario de Estadística Agroalimentaria*, Secretaría General Técnica, Ministerio de Agricultura, Pesca y Alimentación, Madrid

Ministry of Agriculture (2001) *Plan Nacional de Regadíos. Horizonte 2008*, Dirección General de Desarrollo Rural, Subsecretaría, Ministerio de Agricultura, Pesca y Alimentación, Madrid

Ministry of Agriculture (2002, 2006) *Encuesta sobre Superficies y Rendimientos de Cultivos (ESYRCE)*, Secretaría General Técnica, Ministerio de Agricultura, Pesca y Alimentación, Madrid

Ministry of Environment (2006) *Balance de la Política de Agua*, MIMAM, Madrid

Ministry of Environment (2007) *El Uso del Agua en la Economía Española*, Grupo de Análisis Económico del Ministerio de Medio Ambiente, MIMAM, Madrid

Pérez-Pérez, E. (1997) 'El marco legislativo a la gestión del agua de riego', in J. López-Gálvez and J. M. Naredo (eds) *La Gestión del Agua de Riego*, Fundación Argentaria and Visor Distribuciones, Madrid

Sanchis-Ibor, C. (2002) 'La zona regable de la Pedrera (Alicante): La creación de un regadío deficitario (1970–1982)', in *La Directiva Marco del Agua: Realidades y Futuro*, Fundación Nueva Cultura del Agua-Universidad de Sevilla, Universidad Pablo Olavide, Sevilla

Sahuquillo-Herraiz, A. (1984) 'Las aguas subterráneas en España', *Revista El Campo*, no 97, pp17–24

Senent-Alonso, M. (1984) 'Los recursos de agua: Las aguas subterráneas', in *El agua en la Región de Murcia*, Consejo Económico Social de la Región de Murcia, Murcia

12

The Effects of Water Markets, Water Institutions and Prices on the Adoption of Irrigation Technology

Michael D. Young

Introduction

Organisation for Economic Co-operation and Development (OECD) research shows that Australian irrigators have increased water use efficiency at a rate that is at least double the rate observable in other parts of the OECD. In the decade from 1991 to 2001, Australia reduced water use per hectare by 50 per cent while only reducing the area under irrigation by 6 per cent. The purpose of this chapter is to explore some of the reasons why this has occurred and, more specifically, explore the role of technology in helping to bring about this dramatic increase in the technical efficiency of water use.

Figure 12.1 shows the location of irrigation farms in Australia. Each dot represents an area of $1km^2$ where irrigated land use is dominant. The largest concentration of irrigation is in the south-east corner of the country in an area known as the Southern Connected River Murray system. While the majority of irrigators rely on access to water held in storages and released into highly regulated surface water, there are also a significant number of groundwater dependent irrigation systems and a significant number of areas where irrigators divert passing flows into on-farm storages.

By way of background, irrigation water use in Australia is highly regulated and, in order to access water, irrigators must hold a licence to use and an entitlement to access water. Nearly all water use is metered and most water allocations

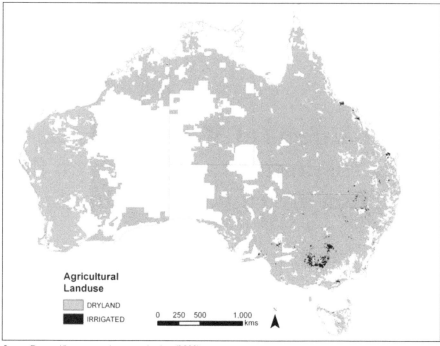

Source: Darren King, personal communication (2008)

Figure 12.1 *Location of irrigation farms in Australia*

are made on a seasonal basis. Formal planning processes are used to determine the amount of water to be allocated to the environment, to towns, to irrigators, etc. Moreover, while considerable progress has been made in the adoption of more efficient irrigation technologies, progress in the resolution of over-allocation problems has been less impressive. Markets facilitate change at the individual level but do not enable changes to the administrative foundations upon which the market is built.

Australia has only recently begun to collect comprehensive data on water use efficiency on a national scale. Nevertheless, the emerging data show that the amount of water applied per farm hectare continues to improve. In 2003, water use was 4.4ML (or 4400m³) per hectare, the next year it fell to 4.3ML per hectare and the year after that it had fallen to 4.2ML per hectare (Table 12.1). Much of this improvement in water use efficiency is due to the adoption of new technologies and the way irrigation and other farmers have been willing to change the mix of crops they grow. As a result of autonomous market-driven structural adjustment processes, the number of Australian farms involved in irrigation has been declining, but the area under irrigation has been increasing. As a result of access to and a willingness to embrace technological opportunities that make money, Australian farmers are skilled at doing more with less.

Irrigated area, irrigation water use and irrigation water application rates

% change in irrigated area 1990-92 to 2001-03		Irrigated area ('000 hectares)		Change in irrigated area ('000 hectares)		% 1990-92 to 2001-03	Change in total agricultural area % 1990-92 to 2001-03	Share of irrigated area in total agricultural area % 2001-03	Share of irrigation water use in total agricultural water use % 2001-03	Irrigation water application rates (Megalitres per hectare of irrigated land)		
		1990-92	2001-03	1990-92 to 2001-03						1990-92	2001-03	% Change
	New Zealand	250	475	225	90	-3	4					
	Belgium	24	40	16	67	2	3	22	0.1	0.2	104	
	France	2150	2632	482	22	-2	9					
	Canada	900	1076	176	20	-3	2	94	3.5	3.6	1	
	Australia	2057	2402	345	17	-6	1	90	8.7	4.3	-50	
	United States	19994	22884	2890	12	-4	5	99	9.4	8.4	-10	
	Sweden	48	54	6	12	-6	2	70	2.1	1.7	-19	
	Spain	3200	3442	242	8	-2	9	100	7.4	7.0	-5	
	OECD	48979	52830	3850	8	-3	4		9.2	8.4	-9	
	EU15	11778	12618	840	7	-3	9		5.6	6.1	8	
	Turkey	3329	3506	177	5	1	9		5.7	8.8	56	
	Greece	1383	1431	48	3	0	17	100	5.5	5.9	7	
	Denmark	433	448	14	3	.5	17	93	0.7	0.4	-48	
	United Kingdom	165	170	5	3	-10	1	9	1.0	0.6	-43	
	Portugal	631	650	19	3	-4	17	100	8.1	9.5	18	
	Mexico	6170	6320	150	2	1	6	97	9.9	8.7	-12	
	Netherlands	560	565	5	1	-3	29	80	0.3	0.1	-59	
	Germany	482	485	3	1	-1	3		3.3	0.3	-91	
	Austria	4	4	0	0	-3	0	5	12.5	2.5	-80	
	Italy	2698	2698	0	0	-1	17	100		7.7		
	Poland	100	100	0	0	-8	0.6	8	3.7	0.9	-77	
	Switzerland	25	25	0	0	-3	2					
	Japan	2846	2641	-205	-7	-8	55	99	20.4	21.3	5	
	Korea	984	880	-104	-11	-12	46		14.3			
	Hungary	205	126	-79	-39	-8	2	21	2.1	1.2	-44	
	Slovak Republic	299	153	-146	-49	0	6	73	0.5	0.4	-31	
	Czech Republic	43	20	-23	-54	0	1	60	0.7	0.6	-21	

% -20   -10   0   10   20

Source: OECD (2008)

Figure 12.2 *Changes in irrigation efficiency in OECD countries*

In order to understand the factors driving this increase in water use efficiency, it is necessary to consider at least four factors. The first is a widespread national commitment to exposing Australian agriculture to international competition. As a result, farmers are forced to continually search for and adopt new, more cost-effective forms of technology. The second factor is a national commitment and public support for government control over water use and, as a result, a strictly enforced licensing programme that limits the quantity of water that may be used in any region. The third factor is an extensive water reform programme, which

Table 12.1 *Changes in water use efficiency in Australia as suggested by data on the water application rate per hectare*

		Year		Change (%)
	2002–03	2003–04	2004–05	
Agricultural establishments irrigating (no)	43,774	40,400	35,244	−19.5
Area irrigated (000ha)	2378	2402	2405	1.1
Volume applied (ML)	10,403,759	10,441,515	10,084,596	−3.1
Application rate (ML/ha)*	4.4	4.3	4.2	−4.5

Note: * Averaged across all irrigated pastures and crops grown in Australia.
Source: Australian Bureau of Statistics (2007)

has its roots in a commitment to competition in the provision of access to and delivery of water to users and, hence, the development of water markets as a means to facilitate change. The fourth factor is the impact of a prolonged drought which has forced all irrigators to work out how to increase their production using less water and whether or not they should use or sell any of the limited amount of water they currently have access to.

Irrigation technology

Produced by the Australian Bureau of Agricultural and Resource Economics, Table 12.2 provides a simple snapshot of the use of technology by irrigators in the production of vegetables in Australia. In Australia, many vegetable farmers are taking courses on chemical use, testing for residues and using computers. Although these data are only for the vegetable industry, they are typical of the way most Australian irrigators approach and use technology.

Investment in 'new' irrigation technology in Australia is widespread. Driven by signals from the water market, increasing water scarcity and increasing information, the frontier of farm technology development is now focused in four areas:

1   evaporation mitigation on farm storages;
2   adaptive irrigation control systems;
3   soil and root zone analysis; and
4   measurement and monitoring systems (Raine and Durack, 2005).

There remains, however, little objective analysis of the collective impact of this broad suite of arrangements from a productivity or technology viewpoint. But we can provide the generalized observation that industries, like the grape industry, have moved quickly to adopt partial root-zone drying and other deficit irrigation practices. Similarly, almost all rice and cotton is only grown on laser-levelled fields.

As a general rule, as the rate of technology adoption increases, water-use efficiency increases. Over the past 25 years, rice yields have risen from 5 to 10 tonnes per hectare (Figure 12.3). In other parts of the irrigation sector, investments have been made in lateral move and centre-pivot irrigation systems on broad-acre irrigation cropping farms. Many irrigation systems operate under pressure and nearly all irrigation water use is metered.

Economic analysis of options is common. A recent National Water Commission study found that:

•   the adoption of laser levelling produced a positive rate of return once the farmer's equity in the investment exceeded 50 per cent;

Table 12.2 *Selected technology-related data summarizing the nature of irrigation in Australia's vegetable industry*

Source of irrigation water, by state, 2005–2006								
Average per farm	NSW %	Vic %	Qld %	SA %	WA %	Tas %	NT %	Aust %
Irrigation scheme	35	35	19	24	23	15	0	26
Groundwater bore	6	24	52	56	35	7	90	31
Diversion from river/stream	1	5	3	0	0	8	10	3
Town water (mains supply)	5	10	7	14	7	2	0	7
Farm storage dam	40	20	17	3	33	67	0	28
Treated or reclaimed water	0	1	0	3	0	0	0	1
Other	3	1	3	0	2	0	0	2

Chemical use, by state, 2005–2006								
Percentage of farms	NSW %	Vic %	Qld %	SA %	WA %	Tas %	NT %	Aust %
Growers undertaking chemical course in the past 3 years	80	78	63	68	57	50	46	68
Growers reducing chemical use per ha in the past 3 years	80	84	89	87	73	87	88	84
Growers testing for chemical residues	31	70	72	73	63	67	34	61

Pests and diseases, by state, 2005–2006								
Percentage of farms	NSW %	Vic %	Qld %	SA %	WA %	Tas %	NT %	Aust %
Farms following a pest and disease monitoring programme	85	90	73	76	64	70	18	77

Education and training, by state, 2005–2006								
Percentage of farms	NSW	Vic	Qld	SA	WA	Tas	NT	Aust
Percentage of growers attending:								
– Conferences	23	35	27	38	37	58	24	34
– Field days	61	76	56	65	62	96	34	67
– TAFE	24	6	3	9	0	12	6	9
– University	0	0	0	0	0	0	0	0
– Workshops	53	49	34	67	63	59	34	50
– Other	14	0	3	7	4	4	0	5

Use of computers in vegetable business, by state, 2005–2006								
Average per farm or percentage of farms	NSW %	Vic %	Qld %	SA %	WA %	Tas %	NT %	Aust %
Percentage of growers using a computer in their business	55	88	58	54	61	76	55	65

Source: Australian Bureau of Agricultural and Resource Economics (2008)

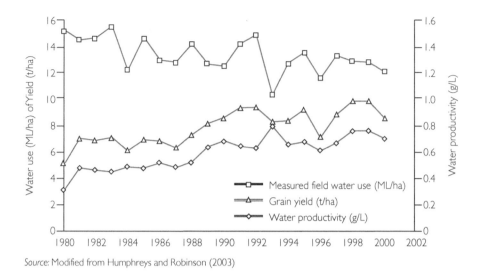

Source: Modified from Humphreys and Robinson (2003)

Figure 12.3 *Trends in rice water productivity, grain yield and field water use in the Murrumbidgee Irrigation Area*

- returns on the farmer's equity were about 45 per cent when the equity was around 50 per cent when one-third (70 hectares) of the farm was irrigated with a centre pivot; and
- returns from the adoption of 70 hectares of subsurface irrigation were 21 per cent when the farmer's equity was at 50 per cent (National Water Commission, 2006).

Off-farm, the main technological frontier is in the development, installation and use of total channel control systems and in catchment-wide planning systems. As virtually all water use by farms is metered in Australia, it is possible to monitor the efficiency of both delivery systems and water use on farms. A considerable amount of comparative data on water use efficiency are available and widely discussed.

Having presented a selective overview, we can now turn to the three main factors driving the adoption of irrigation technology in Australia: open competition, water reform and the extended nature of the current drought.

Open competition

In Australia, producer subsidies for agriculture are in the region of 4 per cent which, by international standards, is extremely low (Figure 12.4). As a result, for many years, Australia has experienced a near continual improvement in agricultural productivity – despite near continual declines in the terms of trade faced by these same farmers (Mullen and Crean, 2007).

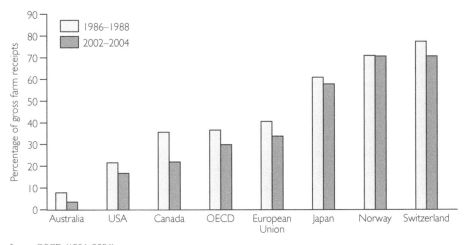

Source: OECD (1986–2004)

Figure 12.4 *Comparison of agricultural subsidies by country measured as the amount of support to agriculture as a percentage of gross farm receipts*

Once again, while no internationally comparable irrigation-industry-specific data are available, there are many data on the effects of international competition on the adoption of technology in Australian agriculture. Nevertheless, as Australia's irrigation industry has been responsible for much of Australia's productivity growth in agriculture, it seems reasonable to assume that the technological benefits of this approach are at least in part due to improvements in irrigation efficiency. In Australia, the irrigation sector is responsible for around 25 per cent of the gross value of agricultural production (Australian Bureau of Statistics, 2006).

One of the interesting consequences of Australia's open, competitive approach to agricultural development has been a continual increase in productivity growth for a long period of time in Australia's specialist crop industries which, as a general rule, tend to be very dependent on access to irrigation water. Productivity growth is usually measured by estimating total factor productivity, defined as 'the ratio of a quantity index of all marketable outputs to the corresponding quantity index of all marketable inputs (Coelli et al, 1998). Total factor productivy (TFP) is a measure, over a data period, of the annual proportional rate of improvement in the technical efficiency with which farmers combine marketable inputs to produce marketable outputs' (Kokic et al, 2006).

Over the past few decades, total factor productivity in Australia has been growing annually by between 2 and 3.8 per cent per annum. On a product by product basis and when corrected for the effects of water availability, the main reason for this general increase in productivity has been a continual market-driven search for and adoption of new, more productive technology. As Kokic et al (2006) observe, 'sustained productivity improvements have long been the engine of growth of Australia's agriculture sector'. Continuously exposed and unprotected from global markets, Australian farmers are forced to pursue more efficient ways of producing more output with less input to offset what are known as declining 'terms of trade'. Unfortunately, terms of trade have been in continual decline for many years. The solution to this riddle is to adopt new technology.

In addition to a competitive farm sector, in recent years, Australia has also invested heavily in the modernization of irrigation supply and delivery systems. This includes the use of meters in the supply of irrigation water to virtually all farmers and, more recently, the introduction of total channel control systems and an incentive-driven search for ways to reduce delivery losses.

Water reform

In the early 1990s, Australia became aware of the fact that while it had been pursuing an open, competitive approach to agricultural development, the way water was managed was an exception to the general rule. And, as a result of this self-criticism, in 1994 the Australian Council of Governments committed Australia to a National Competition Policy (NCP). The main feature of this national policy commitment was a requirement for each state to increase competition and efficiency in the transport, electricity and water sectors that are run (some think inefficiently) by government departments. Government businesses were expected to be become as efficient and responsive to the need for change as other businesses already were. A series of reform milestones were set and it was agreed that any state that failed to meet any of these water reform milestones would be penalized.

Australia is a country that was formed just over 100 years ago through the federation of a number of independent states which, while reliant on the federal government for much of their revenue, are independently responsible for water management.[1] The NCP rule was simple: if a state failed to implement agreed reforms in time, large amounts of money would be withheld from them and, if the situation not quickly rectified, this money would then be shared among the other states. Under NCP, one of the biggest areas for reform was water.

These competition-driven water policy reforms were followed by the introduction of a National Water Initiative in 2004 and, most recently, followed by a National Plan for Water Security (which has yet to take effect).

In addition to these reforms, 1994 was also the year when a new Murray–Darling Basin Agreement was introduced (Young et al, 2006). The Murray–Darling Basin is Australia's largest river system and the place where most of Australia's irrigation water is used. Reform without attention to the Murray–Darling Basin would not be reform at all.

Summarized in more detail in Annex 1, from an irrigation technology perspective, the main features of these reforms included:

- Scarcity – A virtual freeze or cap on the issuance of new irrigation licences in most irrigation areas and the introduction of enforced limits on the amount that any person may extract.
- Trading – A requirement that land titles and licences to take water be separated from one another and made tradable.
- Administrative separation – A requirement that the legal entities responsible for the distribution and supply of water be either privatized or corporatized in a manner that keeps them 'at arm's length' from policy-makers.
- Pricing – A requirement for all irrigation water to be supplied at the full cost of providing this water and maintaining storage and delivery systems.
- Planning – A requirement for the development of formal catchment management plans coupled with regulations that require investment in salinity control and other practices.
- Reduced allocations – A requirement to restore all river systems to sustainable levels of use.

In addition to these policy reforms, investment in irrigation research, catchment planning processes and administration was increased considerably in past decades in a manner that has encouraged the adoption of more efficient irrigation technologies. Despite all these reforms, Australia is still finding it difficult to achieve the balance between the most appropriate amount of water to allocate to users and to the environment. Despite many attempts and the development of many plans to resolve this issue, many systems remain over-allocated in the sense that too little water is being allocated to the environment.

Scarcity and trading

Ultimately, a time arrives when a water resource becomes fully developed and a limit needs to be placed on further use. In much of Australia, it became clear that water resources became fully developed during the 1980s. During this time, a policy transition occurred and states began declaring that water resources were fully developed. Arrangements were then put in place to limit further use and force any further development to occur either through the increase of water use efficiency and/or the trade of water entitlements and allocations from one farm to another. In many areas, the decision to declare a region as 'fully developed' was taken too late and. As a result, a significant number of water resources are now over-allocated. Politically difficult discussions about how best to solve these over-allocation problems are now under way.

The first step in this process was to separate water licences from land titles so that entitlements to access water could be moved from one location to another. Seasonal allocation procedures were also improved so that so-called temporary or seasonal trading of water could commence. Markets for the trade of volumes of water are now well established in Australia and arrangements for the permanent trade of water entitlements from one location to another are

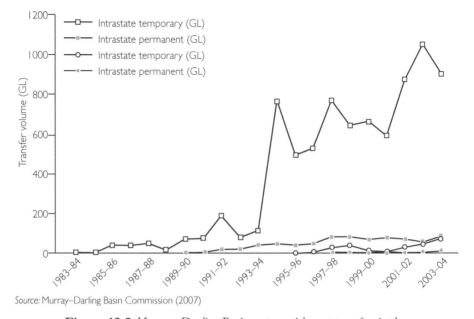

Source: Murray–Darling Basin Commission (2007)

Figure 12.5 *Murray–Darling Basin water entitlement transfers in the Southern Connected Basin*

becoming more sophisticated. So much so that today it is possible to arrange to trade River Murray water over the internet. Water trading is now very much part of the day-to-day operation of irrigation business in most large irrigation systems in Australia that are dependent upon access to regulated water supply systems. This has driven the adoption of much technology. Figure 12.5 shows the rate of growth in water trading in Australia.

In order to reduce the costs of trading water, one Australian innovation that has proved to be very important in facilitating trade has been the unbundling of water rights into at least three separate components. The most common approach to the definition of a water entitlement is to define the access entitlement as a share and then periodically make volumetric allocations to these share entitlement holders. Externalities associated with water use are managed using separate location-specific water use approvals (Young and McColl, 2005).[2] One of the main externalities of concern in Australia is irrigation salinity.

Once a well-developed water market is in place, it has been the Australian experience that water trades to the place where it can be most efficiently used – across space and through time. Introduction and development of these water markets is not without problems. Many mistakes have been made and some are proving extremely hard to rectify. Knowledge about the best way to set up a market is now one of the new technologies that Australia can offer the world. In this regard it is particularly important to define, record and allocate water entitlements and water allocations in a manner that is robust enough to allow markets to operate. It is also important to set up accounting and allocation arrangements

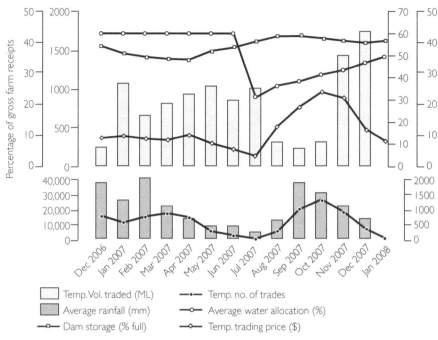

Source: www.waterfind.com.au, accessed 10 January 2008

Figure 12.6 *National Water Index developed by Waterfind showing the nature of relationships between rainfall, storage levels, allocation announcements and price (regional indices are also available and, for decision making, are more useful)*

that are consistent with the way that water flows across and through landscapes. Unless one is careful, it is easy to set up water trading arrangements that cause a region to trade into trouble. Moreover, once problems emerge it is very hard to rectify them.

Figure 12.6 presents data from a National Water Index developed by one of Australia's larger internet-based water trading businesses. In regions like the Southern Connected River Murray system, the water market is now so sophisticated that farmers have a strong incentive and opportunity to manage both within-season and between-season supply risk carefully. They also have a strong incentive to continually decide whether or not they can make more money by selling water to someone else or by using it themselves. In the current drought, dairy farmers have been quick to point out that the presence of this seasonal water market has made it easier for them to survive (Young et al, 2006; Frontier Economics, 2007. One of the main strategies used by these dairy farmers is to continually monitor the cost of buying in feed versus the cost of irrigating pasture. When the price of water rises, they tend to purchase grain from dryland farmers and feed this grain to their cows. When the price of water is low, these dairy farmers tend to grow more irrigated pasture and feed this pasture to their cows.

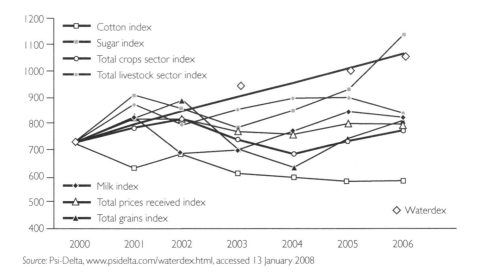

Source: Psi-Delta, www.psidelta.com/waterdex.html, accessed 13 January 2008

Figure 12.7 *Relationship between long-run water price index (Waterdex) and indices of the value of agricultural commodities developed by the Australian Bureau of Agricultural and Resource Economics*

One of the most important lessons from Australian experience with the development of water markets is that as prices increase, interest in technology expands and this interest drives innovation. That is, water markets encourage the adoption of existing technology and also encourage the development of new technology. 'Smart' irrigation technology is a substitute for water.

While prices for Australian agricultural products have been relatively flat in recent times, the presence of smart irrigation technology has meant the price of water has risen as irrigators compete to take greater advantage of the knowledge and opportunities that are emerging (Figure 12.7). It is Australian experience that technology tends to expand production opportunities and, as it does, the value of water in the market rises. As this occurs, the opportunity cost of using water inefficiently rises in a manner that encourages those involved in inefficient use to either change practice or drop out of the system.

Administrative separation and pricing

In parallel with a decision to establish markets for water trading, in 1994 all Australian governments agreed collectively to move the provision of water supply services from departments responsible for the provision and implementation of water policy. Among other things, this policy reform forced those responsible for the provision of water to irrigators to operate on what is known as a 'level playing field' with private enterprise. In some states, this has been achieved through the establishment of water-supply companies owned by the government as a single shareholder. In other states, especially in New South

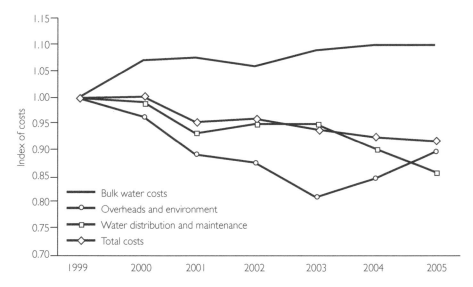

Figure 12.8 *Index of water supply and delivery costs in real terms since privatization of Murrumbidgee Irrigation*

Wales, ownership was transferred to corporations that are owned and operated by all the irrigators in each area.

The first consequence of this decision to transfer control and responsibility for management of irrigation systems to irrigators has been an immediate shift to a regime that forces these companies to pay for the full cost of maintaining 'their' irrigation system and supplying water to themselves. Interestingly, the transfer of control of the distribution system to irrigators in NSW has been welcomed by irrigators and, to date, proved to be extremely successful in reducing the cost of delivering water. In the case of Murrumbidgee Irrigation, for example, while the cost of bulk water charges imposed by the state has risen in real terms, in the first six years of operation, the real total cost of supplying water to irrigators in the Murrumbidgee fell (Figure 12.8). Once irrigators had control of the system, they had no one to complain to apart from themselves so they decided to cut costs.

As a direct result of these administrative reforms, two noticeable changes are now under way. The first change is the development and adoption of total channel control technologies. The second is the emergence of a willingness of the entire irrigation community to carefully consider and evaluate opportunities for system modernization. (Much of this work has also been supported by Australia's National Water Commission under its Water Smart Australia programme. Eight case studies summarizing some of the success stories are described by the National Water Commission (2006).)

One of the features of transferring control of local supply systems to irrigators was a decision to transfer ownership of delivery losses to them by issuing their supply company or cooperative a bulk water entitlement that included an allowance for delivery losses. Once such a bulk water entitlement is issued, water-supply companies have an incentive to search for ways to reduce delivery losses as any savings made can be passed on to irrigators in the form of increased allocations. In the case of Colleambally Irrigation, for example, this arrangement provided sufficient incentive for irrigators to work with Rubicon Pty Ltd to develop total channel control systems that have dramatically increased the efficiency of their delivery systems. This was achieved by combining the use of automated and radio-controlled flume gates, internet-based water-ordering systems and flow optimization programmes to coordinate the delivery of water to all the farms in any part of the supply system. The result is an open channel control system that boasts 91 per cent delivery efficiency. As a result, last year the impact of drought on irrigators in the Colleambally Irrigation area was much less than it otherwise would have been. Other water supply companies in the Southern Connected River Murray system are now in the process of installing similar total channel control systems.

The second noticeable feature of the transfer of control and management responsibility to irrigators has been the emergence of a willingness to consider alternative charging rules and system configurations. In the Pyramid Hill Boort area in Victoria, for example, irrigators have started asking whether or not there is a more efficient way to charge for access to water and to reconfigure their system. Figure 12.9 (Plate 13) summarizes the nature of the data now being shared with all the irrigators in the system. It shows on a channel-by-channel basis how much it costs to deliver water to each part of the system. An important discussion is now under way to determine whether or not the practice of 'postage stamp' pricing should continue. Postage stamp pricing involves the setting of the same supply charge per megalitre irrespective of how much it costs to get the water to the field. Options under consideration include charging for the actual cost of delivery and/or closure of those parts of the system that are expensive to maintain. Supply companies have also started to charge irrigators an exit fee if they decide to transfer their entitlement and delivery entitlement out of the district.

Planning

From a technology perspective, the next part of Australia's water reform programme has been the appointment of catchment management boards that, within the constraints set by state-wide policies, are responsible for developing land use and water sharing rules. Once made, these plans become legally binding documents that can be changed only through quite complex consultative processes. From a technology perspective, one of the argued benefits of this formal planning approach is that it is supposed to provide the certainty necessary to allow irrigators to invest with confidence.

While the degree to which formal planning approaches increase certainty and through this increase the rate of technology adoption is difficult to quantify,

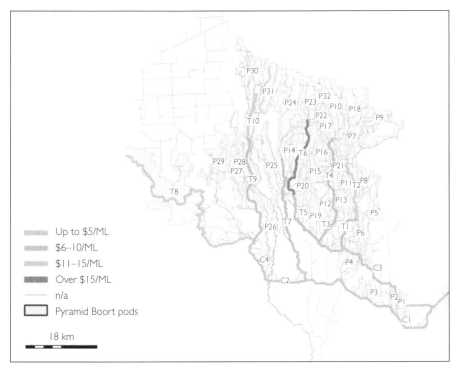

Note: See Plate 13 for a colour version.
Source: Goulburn Murray Irrigation, personal communication

Figure 12.9 *Annualized costs of water delivery to farms in the Pyramid Hill Boort Irrigation Area*

economic theory would suggest that providing the plans are sound, the result should be more investment.

One of the spin-off consequences of this approach, however, has been the development of sophisticated approaches to the management of adverse environmental impacts of irrigating land. As a result of these processes, in many areas it is now only possible to gain approval to irrigate a previously unirrigated area of land if a commitment is made to use the 'best' available technology. In addition, in places like the River Murray in South Australia, existing irrigators are being required to develop and submit land and water management plans which require each irrigator to demonstrate how they will meet agreed water use efficiency benchmarks.

Reduced allocations

Another problem that water reform in much of Australia has had to deal with is the over-allocation and over-use of water. In Australia, over-use is said to be occurring whenever the volume of water being used is greater than the amount of use (extraction) that the system can sustain given current conditions. Over-allocation is said to be present when the sum of all entitlements to access water is greater than could be sustained if all entitlements were exercised. Among other things, over-allocation is aggravated by the fact that increases in water use efficiency tend to reduce the volume of water that returns to rivers and associated aquifers (Young and McColl, 2005).

Regrettably, as it introduced trading, Australia has found it extremely difficult to manage over-allocation problems. Whenever water trading is introduced, the first water entitlements and allocations to be used typically are entitlements which before that time had not been sold. In many Australian systems, this resulted in an increase in the total amount of water being used. In the first review of the pilot interstate water trading trial in the Murray River system, for example, Young et al (2000) found that virtually all of the early interstate water entitlement trades that occurred involved the sale of water that was not being used. As a result, the first thing that the introduction of water trading did was to increase the total amount of water used in any system. In retrospect, the golden rule in implementing any system that enables water to be traded is to begin by putting in place institutional arrangements that are robust enough to prevent over-use from occurring. And once a system is fully developed, a regime must always be put in place in a manner that stops over-allocation from occurring. One of the simplest ways of introducing trading without encountering over-allocation problems is to define each entitlement as a share of the amount of water use that can be sustained and then reduce allocations per share as water is allocated. Another way is to cancel all or part of the unused proportion of each allocation – many variants and combinations of these two approaches are possible.

Increasing water scarcity – drought

Since 2002, much of eastern and southern Australia has suffered a drought that is of much longer duration than any that has been experienced in the last half of the 20th century. The result has been the emergence of a situation where allocations to high security water entitlements have had to be reduced significantly. At the time of writing, allocations to high security entitlement holders in South Australia, for example, are at 32 per cent of the allocation these irrigators have always received until this current drought. One of the most noticeable consequences has been a decision to abandon mature plantations and plantations that have high delivery costs.

Once again, from a technology viewpoint and while the impact of a long drought is something that no one would recommend, this adverse experience is forcing all irrigators to carefully review all their strategies and, in particular, search for ways to use water more efficiently. As a result of this experience and the increasing number of publications warning irrigators to expect adverse climate change, Australia is now actively reconsidering the way that water entitlements are defined and allocations managed. Emerging, but still contested, lessons include recognition of the importance of:

- defining entitlements as shares rather than in volumetric terms;
- allowing irrigators rather than river managers and government administrators to decide how much water to carry forward from year to year;
- giving first allocation priority to putting aside enough water to account for system losses and then giving the environment and all other water users a share of the available water in a manner that forces all users to manage interseasonal risk;
- accounting for and regulating all unmetered forms of water use;
- avoiding the temptation for governments to selectively interfere in the water market; and
- avoiding the temptation during times of stress to selectively subsidize less efficient irrigators (Young and McColl, 2008).

Conclusion and policy implications

The overall question that this chapter seeks to explore is whether or not the introduction of water trading, water pricing and administrative separation policies is favourable for the adoption of technology. Australian experience suggests that the answer to this question is very much in the affirmative. Moreover, there is nothing in the Australian experience which suggests that these lessons are immediately transferable to other countries.

As a general rule, it has been the Australian experience that water trading encourages the transfer of water to areas where it is more likely to be associated with the adoption of new technology. Moreover, the revenue generated from the sale of water to others is often used to raise the capital necessary to invest in new technology. It is also creates a strong incentive, especially in droughts, for irrigators to abandon old technology and inefficiently configured irrigation systems.

Pricing reforms and administrative reforms appear to have a similar effect. In particular, provided governance issues are sorted out, the separation of responsibility for policy from the management and delivery of water to irrigators is likely to encourage the adoption of new delivery technologies and management systems.

Consideration of the merits of transferring total ownership and responsibility for the management of delivery systems is an approach that has considerable merit, especially when the savings for investments in improved system efficiency are made available to irrigators. When estimating the nature and extent of savings, however, it is important not to confuse reductions in returns to the river and to aquifers with evaporative savings. In most cases, reductions in supply system leakage and reductions in return flows are not savings at the system level.

The above strong recommendation for the introduction of water trading as a means to promote the adoption of more efficient technology is accompanied by a parallel recommendation and important qualification. It is critical that both the specification of water entitlement and allocation systems traded are robust. In particular, care must be taken to specify entitlements and allocations in a manner that is consistent with the way that water flows across landscapes and through land.

While generally recommending that countries use markets as a means to stimulate the adoption of more water-use-efficient technologies, one final caveat must be made. Reform sequencing is important. Markets rely on the presence of fully specified water entitlement and allocations systems. They also require national and regional commitment to enforcement.

Annex 1: A history of High Level Water Policy Reform in Australia[3]

The history of High Level Water Policy Reform (HLWPR) that affects irrigation investment and practice in Australia is complex and its nature is difficult to understand. A variety of intergovernmental, Murray–Darling basin-wide and state-level processes have been undertaken. In the past decade and driven by Council of Australian Government (COAG) decisions, the main national building blocks were:

- COAG 1994 – as part of an NCP agenda all Australian governments agreed to introduce policies that would improve water use and management across the nation by introducing reforms that would encourage water to be used in areas where it would create the greatest value.[4] State jurisdictions have responsibility for implementing the COAG Water Reform Framework for the Australian water industry. Payments are made to the states on the delivery of key reform milestones.
- NCC 1995 – Governments agreed to establish a National Competition Council (NCC) that would audit progress in implementing the COAG 1994 agreement and, using a tranche payment system, make a proportion of transfer payments from the Commonwealth to states conditional upon meeting performance targets set out in the COAG 1994 agreement.
- NAP 2000 – National Action Plan for Salinity and Water Quality.
- National Water Initiative (NWI) 2004 – A blueprint for the next decade of reform of Australia's water management which was signed by the Commonwealth and most state and territory governments.

In addition to these high level policy reforms, the Commonwealth government introduced four programmes designed to assist states, communities, businesses and individuals to invest in the restoration and protection of Australia's natural resources. The four programmes were:

- The Natural Heritage Trust (NHT) was set up by the Commonwealth government in 1997 to help restore and conserve Australia's environment and natural resources. A AS$3 billion fund was established to provide grants to community groups and organizations for environmental and natural resource management projects.
- The National Action Plan for Salinity and Water Quality, endorsed by COAG in 2000, provided a significant funding package of AS$1.4 billion to tackle two major natural resource management issues facing Australia's rural industries, regional communities and unique environment through working with people in communities to find local solutions for local problems.
- A AS$2 billion Australian Government Water Fund.
- The Commonwealth government joined with New South Wales, Victoria, South Australia and Australian Capital Territory in a AS$500 million investment to address the declining health of the Murray–Darling River system through the Living Murray initiative.

In parallel with these national reform processes a number of independent reforms were being implemented by the Murray–Darling Basin Commission. These included:

- 1994 – a new Murray–Darling Basin Agreement;
- 1995 – introduction of a 'cap' that limits the volume of surface water that may be extracted from the Murray–Darling Basin system in any year;[5]
- 1998 – introduction of a Pilot Interstate Water Trading Trial along the River Murray (between Nyah and the Barrages at the river mouth);
- 2001 – adoption of a Salinity and Drainage Strategy; and
- 2002 – development of the Living Murray process.

Another feature, common to all states, has been the preparation of new water legislation. In most cases, a number of further amendments to these new acts have been necessary. The approaches used to implement these processes differ considerably. For example, following an extensive consultation process involving a green and then a white paper, Victoria has recently introduced a totally new act that will change the way water entitlements are defined and water is managed across the state. Most recently, New South Wales has used a ministerial statement to accompany a set of amendments that enable water entitlements to be defined using a unit share system.

Perhaps the most dominant of all high level water policy reforms that have ever been made by COAG is to make receipt of competition payments under the NCP conditional upon states meeting a number of water reform targets. Table

Table 12.3 *Annual NCP payments* received by four case study states (A$ million)*

State	Year								
	97–98	98–99	99–00	00–01	01–02	02–03	03–04	04–05	05–06
NSW	126.5	138.7	148.6	155.9	242.5	251.8	203.5	233.6	292.5
Vic	92.8	102.0	109.2	114.7	179.6	182.4	178.7	201.6	197.9
Qld	74.2	81.6	81.5	73.0	147.9	138.9	87.9	143.3	178.7
SA	34.3	38.4	34.5	35.9	55.7	57.1	40.7	50.4	54.3

Note: * These estimates are subject to periodic revision as new consumer price index and population data become available. Consequently, the dollar estimates reported here may differ slightly from the actual payments and penalties determined by the Australian Government in response to the NCC's recommendations.
Source: www.ncc.gov.au/articleZone.asp?articleZoneID=40#Article-93, accessed 13 January 2008

12.3 summarizes the nature of these payments for the period from 1997/1998 to 2005/2006. Payments to each state reflect the relative size of their economy. The amounts of money are large and in many cases the water reform targets required significant changes to existing administrative and legislative arrangements.

In 2004/2005, A$26 million (10 per cent) of New South Wales' competition payments were suspended because the NCC was of the opinion that New South Wales had failed to adequately demonstrate to 'satisfy the COAG obligation to provide appropriate allocations of water to the environment' (NCC, 2004).

In 2007, the Commonwealth government announced a National Plan for Water Security and then introduced a new Water Act that is expected to result in the establishment of an Independent Murray–Darling Basin Authority and give a new Commonwealth government overall responsibility for management of the Murray–Darling Basin. As at the time of writing, states have not yet agreed to this new approach.

Notes

1   Australia is in the middle of a process that may result in the transfer of responsibility for the management of Murray–Darling Basin water resources to the federal government. The intention is to replace the Murray–Darling Basin Commission with a new independent Murray–Darling Basin Authority.

2   For a discussion on the management of externalities in Australia see the parallel paper on Australian approaches to the management of non-point sources of pollution (Chapter 6).

3   Adapted from Young et al (2006).

4   COAG adopted the recommendations of the COAG report in April 1995 and, in 1997, the Prime Minister confirmed that the COAG was to define a reform process for water management in Australian states and territories. The resultant framework embraces pricing reform based on consumption-based pricing and full-cost recovery, the reduction or elimination of cross-subsidies and making subsidies transparent. It also involves the clarification of property rights, the allocation of water to the environment, the adoption

of improved entitlement and allocation trading arrangements, institutional reform and expanded public consultation and participation.
5   Diversions refer to water that is diverted or taken from the river. Diversions include water supplied to irrigators for agriculture and supplied to satisfy stock, domestic and urban needs.

References

Australian Bureau of Agricultural and Resource Economics (2008), Canberra. www.abareco-nomics.com/interactive/vegegrowing/excel/other.xls, accessed 13 January 2008

Australian Bureau of Statistics (2006) *Water Account, Australia, 2004-05*, Publication 4610.0. Australian Bureau of Statistics, Canberra

Australian Bureau of Statistics (2007) *Water Use on Australian Farms*, Publication 4618.0. Australian Bureau of Statistics, Canberra

Coelli, T., Rao, D. and Battese, G. E. (1998) *An Introduction to Efficiency and Productivity Analysis*, Kluwer Academic Publishers, London.

Frontier Economics (2007) *The Economic and Social Impacts of Water Trading*, Report prepared in association with Tim Cummins and Associates, Dr Alistair Watson, and Dr Elaine Barclay and Dr Ian Reeve of the Institute for Rural Futures, University of New England for the National Water Commission, Canberra, available at www.nwc.gov.au/publications/docs/EconomicSocialWaterTrade.pdf

Humphreys, E. and Robinson, D. (2003) 'Increasing water productivity in irrigated rice systems in Australia: Institutions and policies', in T. W. Mew, D. S. Brar, S. Peng, D. Dawe and B. Hardy (eds) *Rice Science: Innovations and Impacts for Livelihood*, Proceedings of the International Rice Research Conference, 16–19 September 2002, Beijing, China, pp885–900

Kokic, P., Davidson, A. and Boero Rodriguez, V. (2006) 'Australian growth industry: Factors influencing productivity growth', *Australian Commodities*, vol 13, no 4, pp705–712

Mullen, J. D. and Crean, J. (2007) *Productivity Growth in Australian Agriculture: Trends, Sources, Performance,* Australian Farm Institute, Sydney

NCC (National Competition Council) (2004) *Assessment of Governments' Progress in Implementing the National Competition Policy and Related Reforms (Volume Two: Water)*, Melbourne, October

National Water Commission (2006) 'Investing in irrigation: Achieving efficiency and sustain-ability case studies', www.nwc.gov.au/agwf/docs/Irrigation%20Case%20Studies%20online%20PDF.pdf, accessed 13 January 2008

OECD (1986–2004) 'Producer and consumer support estimates', OECD Database, www.oecd.org/document/54/0,2340,en_2649_33727_35009718_1_1_1_1,00.html

OECD (2008) 'Environmental performance of OECD agriculture since 1990', Organisation for Economic Co-operation and Development, Paris, www.oecd.org/agr/env/indicators.htm

Raine, S. R. and Durack, M. K. (2005) 'The future direction of on-farm irrigation technolo-gies and practice research', Paper presented to the Cotton Production Seminar, August 2005

Young, M. D. and McColl, J. C. (2005) 'Defining tradable water entitlements and allocations: A robust system', *Canadian Water Resources Journal*, vol 30, no 1, pp65–72

Young, M. D. and McColl, J. C. (2008) 'A future-proofed basin: A new water management regime for the Murray-Darling Basin', The University of Adelaide, Adelaide, www.myoung.net.au

Young, M., Hatton MacDonald, D., Stringer, R. and Bjornlund, H. (2000) *Inter-State Water Trading: A 2-year Review,* Policy and Economic Research Unit, CSIRO Land and Water, Adelaide, Australia

Young, M. D., Shi, T. and McIntyre, W. (2006) *Informing Reform: Scoping the Affects, Effects and Effectiveness of High Level Water Policy Reforms on Irrigation Investment and Practice in Four Irrigation Areas,* Collaborative Research Centre for Irrigation Futures, Adelaide Technical Report No 02/06, available at www.irrigationfutures.org.au/newsDownload.asp?ID=295&doc=CRCIF-TR-0206-col.pdf

Index

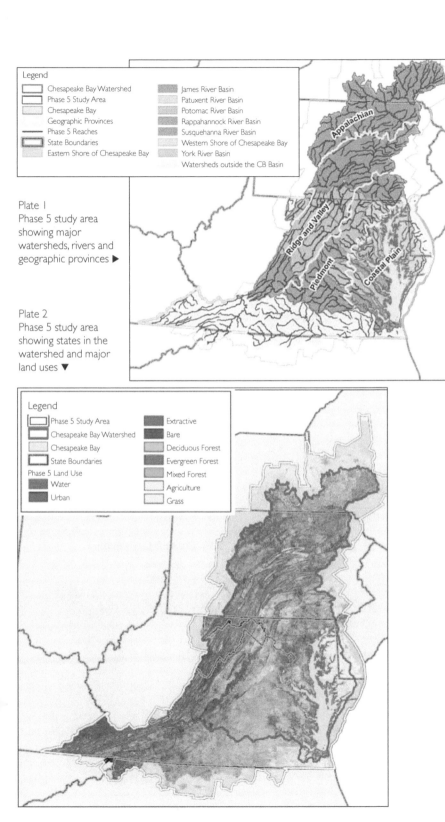

Legend

- Chesapeake Bay Watershed
- Phase 5 Study Area
- Chesapeake Bay
- Geographic Provinces
- Phase 5 Reaches
- State Boundaries
- Eastern Shore of Chesapeake Bay
- James River Basin
- Patuxent River Basin
- Potomac River Basin
- Rappahannock River Basin
- Susquehanna River Basin
- Western Shore of Chesapeake Bay
- York River Basin
- Watersheds outside the CB Basin

Plate 1
Phase 5 study area showing major watersheds, rivers and geographic provinces ▶

Plate 2
Phase 5 study area showing states in the watershed and major land uses ▼

Legend

- Phase 5 Study Area
- Chesapeake Bay Watershed
- Chesapeake Bay
- State Boundaries

Phase 5 Land Use
- Water
- Urban
- Extractive
- Bare
- Deciduous Forest
- Evergreen Forest
- Mixed Forest
- Agriculture
- Grass

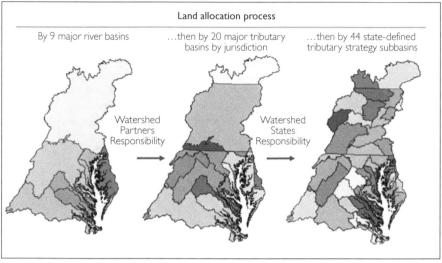

▲ Plate 3

The process used to allocate nutrient and sediment load reductions and caps to first the major basins, then the state-basins, then sub-basins within the state-basin

Plate 4

Comparison of previous Watershed Model phase and Phase 5 segmentation and calibration stations ▼

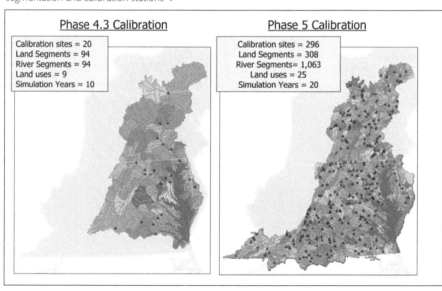

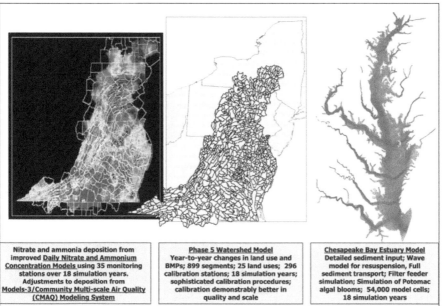

Nitrate and ammonia deposition from improved Daily Nitrate and Ammonium Concentration Models using 35 monitoring stations over 18 simulation years. Adjustments to deposition from Models-3/Community Multi-scale Air Quality (CMAQ) Modeling System
Phase 5 Watershed Model
Year-to-year changes in land use and BMPs; 899 segments; 25 land uses; 296 calibration stations; 18 simulation years; sophisticated calibration procedures; calibration demonstrably better in quality and scale
Chesapeake Bay Estuary Model
Detailed sediment input; Wave model for resuspension, Full sediment transport; Filter feeder simulation; Simulation of Potomac algal blooms; 54,000 model cells; 18 simulation years

▲ Plate 5 Overview of the Chesapeake integrated models of the airshed, watershed and estuary

NY20
NY08
NY67
NY65
NY10
NY68
PA29
PA18
NY99
PA72
PA15
PA42
NJ99
PA00
MD03
WV18
WV99
MD13
DE02
WV05
VA28
DE99
MD18
WV04
VA00
MD15
VA13
KY22
VA24
NC03
NC34
NC41
NC36
NC35

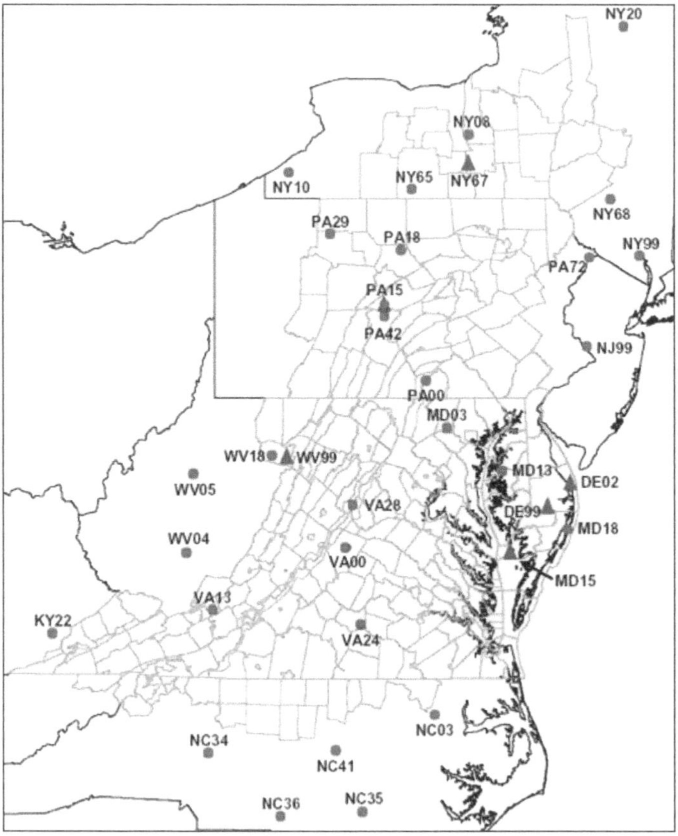

Plate 6 Atmospheric deposition monitoring stations used in the airshed regression model ▶

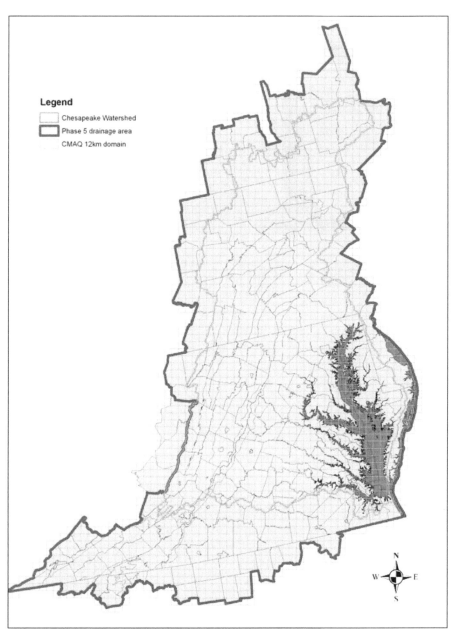

Legend
☐ Chesapeake Watershed
☐ Phase 5 drainage area
☐ CMAQ 12km domain

▲ Plate 7
The 12km CMAQ model grid over the
Chesapeake Bay basin used
for Phase 5 Model applications

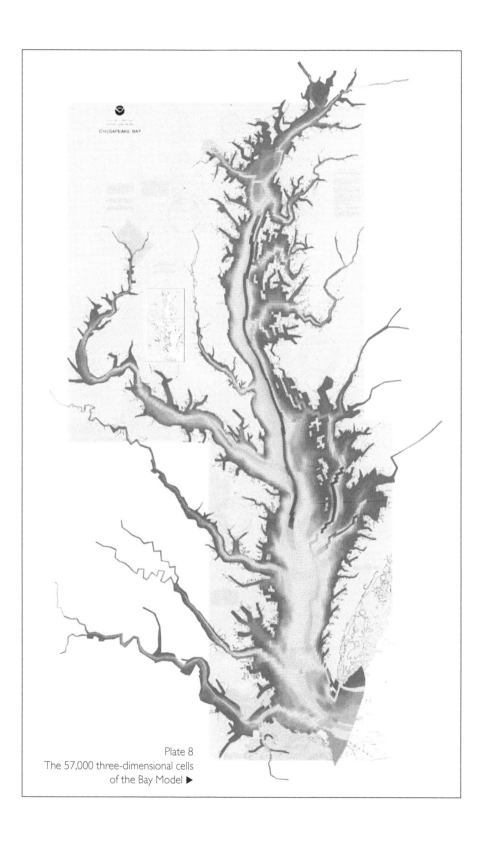

Plate 8
The 57,000 three-dimensional cells
of the Bay Model ▶

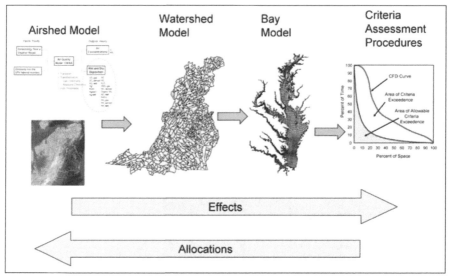

▲ Plate 9
The CBP integrated models of the airshed,
watershed, estuary water quality and
sediment transport, key living resources,
and climate change

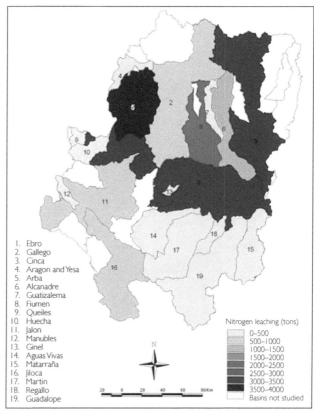

Plate 10
Yearly nitrogen emission
loads in the middle Ebro
basin (t N-NO$_3^-$) ▶

1. Ebro
2. Gallego
3. Cinca
4. Aragon and Yesa
5. Arba
6. Alcanadre
7. Guatizalema
8. Fiumen
9. Queiles
10. Huecha
11. Jalon
12. Manubles
13. Ginel
14. Aguas Vivas
15. Matarraña
16. Jiloca
17. Martin
18. Regallo
19. Guadalope

Nitrogen leaching (tons)
0–500
500–1000
1000–1500
1500–2000
2000–2500
2500–3000
3000–3500
3500–4000
Basins not studied

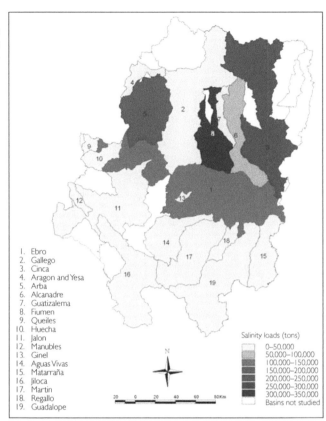

Plate 11
Yearly salinity emission
loads in the middle
Ebro basin (t) ▶

1. Ebro
2. Gallego
3. Cinca
4. Aragon and Yesa
5. Arba
6. Alcanadre
7. Guatizalema
8. Fiumen
9. Queiles
10. Huecha
11. Jalon
12. Manubles
13. Ginel
14. Aguas Vivas
15. Matarraña
16. Jiloca
17. Martin
18. Regallo
19. Guadalope

Salinity loads (tons)
0–50,000
50,000–100,000
100,000–150,000
150,000–200,000
200,000–250,000
250,000–300,000
300,000–350,000
Basins not studied

N

20 0 20 40 60 80 Km

▼ Plate 12
River basin authorities in
Spain

● Planning zone
River basin

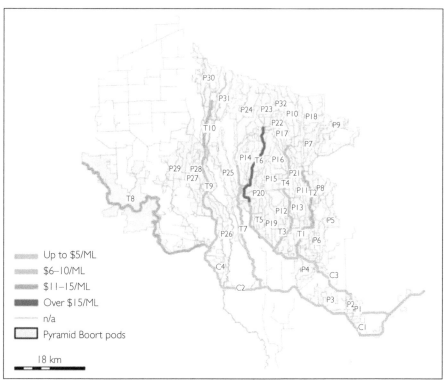

▲ Plate 13
Annualized costs of water delivery to farms
in the Pyramid Boort Irrigation Area

*For Product Safety Concerns and Information please contact
our EU representative GPSR@taylorandfrancis.com Taylor & Francis
Verlag GmbH, Kaufingerstraße 24, 80331 München, Germany*

T - #0093 - 270225 - C0 - 234/156/17 - PB - 9780415849395 - Gloss Lamination